Mahjoub Jabli
Nouha Sebeia

Natural resource exploitation

Mahjoub Jabli
Nouha Sebeia

Natural resource exploitation

Dyeing, bleaching and nanotechnology

ScienciaScripts

Imprint

Any brand names and product names mentioned in this book are subject to trademark, brand or patent protection and are trademarks or registered trademarks of their respective holders. The use of brand names, product names, common names, trade names, product descriptions etc. even without a particular marking in this work is in no way to be construed to mean that such names may be regarded as unrestricted in respect of trademark and brand protection legislation and could thus be used by anyone.

Cover image: www.ingimage.com

This book is a translation from the original published under ISBN 978-620-6-70614-4.

Publisher:
Sciencia Scripts
is a trademark of
Dodo Books Indian Ocean Ltd. and OmniScriptum S.R.L publishing group

120 High Road, East Finchley, London, N2 9ED, United Kingdom
Str. Armeneasca 28/1, office 1, Chisinau MD-2012, Republic of Moldova, Europe
Printed at: see last page
ISBN: 978-620-7-70680-8

Contents

GENERAL INTRODUCTION

General introduction

The treatment of textile materials generally involves pre-treatment, dyeing, printing and finishing. All these stages consume various chemicals and large quantities of water and energy, and produce a large volume of wastewater containing undesirable chemicals. In the field of dyeing, efforts have been made to improve dyeing processes through new modifications to the textile surface or to replace the traditional dyeing process with new environmentally-friendly methods. The application of natural raw materials and renewable sources such as natural dyes has also been considered to reduce the environmental impact of textile dyeing processes. Amid growing environmental and health concerns, non-toxic and environmentally friendly natural dyes have re-emerged as a viable potential 'green chemistry' option as an alternative to some extent to synthetic dyes. The recent resurgence of research and development in the production and application of natural dyes is due to the growing popularity of a more natural lifestyle based on naturally sustainable products. Natural flora and fauna are full of colours that fascinate and attract human beings to a vast portfolio of possibilities.

Water remains one of the most important basic necessities of life on our planet. It covers more than 71% of the Earth's surface and accounts for around 70% of its mass. Water is the only substance on earth that exists naturally in three physical states. The need for pure water cannot be ignored. Unfortunately, over the last few decades, unbridled population expansion and industrial invasion of textiles and other products has led to enormous pressure on limited water resources [1]. In fact, many organic and inorganic pollutants are discharged into the aquatic environment, causing dramatic deterioration in water quality, particularly metals and synthetic dyes, which are carcinogenic and resistant to light and oxidising agents.

Water pollution, a crucial issue, has become a truly global concern. First of all, it is important to understand its nature, its main sources and its harmful effects on human health and the ecosphere. Generally speaking, pollution is environmental contamination that poses acute threats to various living organisms. This pollution is mainly classified into five categories: water, air, soil, biological and toxic.

Many industries are major producers of wastewater loaded with a wide variety of pollutants. These pollutants include dyes, heavy metals, phenols, pesticides, insecticides, drugs, etc. The discharge of these contaminated effluents into the aquatic environment has a huge impact on our biosphere. It has immunogenic, teratogenic, mutagenic and carcinogenic characteristics [2]. It is well known that 70-80% of all cases of disease in developing countries are linked to water pollution [3]. Discharging wastewater into the aquatic environment without treatment causes numerous environmental and health problems for flora, fauna and human health [4].

Because of the impact of dyes and heavy metals on various organisms, the demand for water that purifies them is becoming more and more important. This is a major issue for industry, the environment and health. Physical, chemical and biological treatment techniques have been used to purify the aquatic system. The physical and chemical processes include coagulation-flocculation [5], precipitation [6], ion exchange [7], photo-oxidation [8], electrocoagulation [9], electrodialysis [10], membrane separation [11], ultrafiltration [12], reverse osmosis [13], and adsorption [14]. Biological processes include phyto-extraction [15] and biological degradation [16]. However, most of these processes have a number of limitations, such as high cost, inefficiency at low pollutant concentrations, high consumption of time, reagents, energy and chemicals, and formation of sewage sludge [17]. It is therefore necessary to opt for an alternative, greener and cheaper technology for the effective removal of various contaminants from water. The adsorption process is one of the most cost-effective methods for removing various types of contaminants, particularly organic pollutants (pesticides, dyes, phenolic compounds, etc.) and heavy metals. In addition, researching and developing new natural adsorbents that are abundant, cost-effective and efficient for treating ecosystems is a major challenge. Bio-adsorbents such as biopolymers, solid agricultural waste, algae, etc. have shown promising cost-effectiveness for trapping pollutants. Our country, Tunisia, has a rich and fortuitous plant and animal biomass. This biomass can be used for a wide range of applications, including medicine, drug synthesis, industry, food, dyeing and waste treatment.

To alleviate all the dangers that have arisen as a result of the excessive use of synthetic dyes in dyeing and contaminated water, in this MayuscriLaetude the dyeing properties of some textile materials using new biological extracts extracted from Tunisian fauna, in particular walnut and malt fungi.Then, cellulosic matiëres will be printed using colloidal metal nanoparticles obtained by the reduction of metals by biological aqueous extracts. The biological activity of the printed fabrics will be studied by testing several strains of bacteria.This first part will in fact contribute to the valorisation of natural dyes for the development of ecological dyeing and printing processes in the field of textiles and replace the usual methods using synthetic dyestuff.In the second part, we will study the adsorption process of textile dyes as examples of textile pollutants (methylene blue for example) by raw and modified biomaterials. These studies will be carried out as a function of several parameters (the equilibrium kinetics, the effect of pH, temperature, and that of the initial concentration of the pollutant material) in order to determine the optimum adsorption conditions and to contribute to the understanding of the behaviour of the pollutants at the interface between the polymer particles and the aqueous solution.The aim of this part will be to examine the effectiveness of new biomaterials in trapping cationic dyes. For example, almond waste, date waste, silica, shrimp shell waste, etc. will be used. In the final chapter, we will synthesise new nanoparticles from lignin extracted from poplar fibres and aqueous extracts from oleander plants, pergularia and malta mushrooms, and we will study their physico-chemical characteristics, and their uses also for degrading textile dyestuffs. Finally, we will draw a general conclusion and outline a number of prospects that will serve as starting points for other lines of research. It will therefore be necessary to develop other new adsorbents not only for the degradation of synthetic dyestuffs, but also for the treatment of water contaminated by other types of pollutants such as metals, polyphenols, pesticides, etc.

Chapter I

Extraction of dyestuffs from biomass for dyeing and printing applications

Introduction

Natural dyes, extracted from the roots, stems, leaves and flowers of plants, are still ecological substitutes for synthetic dyes. Their non-toxic effect and biodegradable nature can be explained by the activity of the extracts they contain, such as essential oils, flavonoids, phenols, etc. Their many functional properties could justify the continued progress in their use in a wide range of fields.

In this chapter, we are interested in exploiting new dyestuffs derived from local plant biomass for dyeing and printing. Firstly, we evaluated the dyeing power of some textile fibres using dyestuffs derived from walnut plants and malta mushrooms. Then, cotton fabrics were printed using 2%, 5%, 8% and 10% colloidal metal nanoparticles obtained by reduction of copper, magnesium and nickel salts by an aqueous biological extract of malt fungi. The values of the colouring strengths and colorimetric coordinates will be measured in order to perceive the colour of the coloured material. Resistance to washing, rubbing, light and perspiration will be described. The biological activity of the printed fabrics will be studied, mainly by testing strains of *Staphylococcus aureus*, *Salmonella typhi* and *Candida albicans*.

I. Reminder: colour parameters

Light or colour is defined as any radiation of wavelengths between 400 and 700 nm falling on the retina (1 nm = 10-9 m = 1 billionth of a metre). The colours in the spectrum for each of these wavelengths are violet from 400 to 430 nm, blue from 430 to 485 nm, green from 485 to 570 nm, yellow from 570 to 585 nm, orange from 585 to 610 nm and red above 610 nm. A colorimetric study consists of determining the CIELAB colorimetric parameters (L*, a*, b* and c*) and the degree of absorption (K/S).

In our study, colorimetric measurements were carried out using a spectrophotometer (Data Color 650®, USA). It is interesting to work in a polar co-ordinate space, allowing the light stimulus to be coded using the notions of intensity (L*), 'saturation' and 'hue'. The notion of hue can be approximated by the hue angle (h). The notion of the degree of colouration, i.e. saturation, can be approached by the magnitude (c*). Hue represents what we call "colour" in everyday language.

It distinguishes objects by the qualifiers red, green, blue, etc. irrespective of their luminosity. Saturation seeks to express the degree to which a colour, initially pure (saturated), can be "washed" with white.

It is also important to measure colour intensity (also known as colour strength) using the KUBELKA and MUNK formula:

$$K/S = (1-R)/2R^2$$

For the remission coefficient R of a thick, opaque sample, with the constants K and S, KUBELKA and MUNK obtained, under certain well-determined conditions and at wavelength 1, the ratio K/S = (1-R)2/2R

II. The solidity of dyes

In order to assess the dye fastnesses of the dyed samples, we used the following international standards: ISO 105-C06 for the estimation of fastness to washing, ISO 105-X12 for the estimation of fastness to rubbing, and ISO 105-B02 for the estimation of fastness to light.

A Crockmetre, a cotton test cloth measuring 5 cm x 5 cm, two test tubes each measuring 14 cm x 5 cm and a scale of white were used to measure the friction resistance. When dry, the cotton cloth was placed on the end of the Crockmetre's ankle and moved back and forth 10 times in 10 seconds. When wet, the same test is performed except that the friction cloth (cotton cloth) must be wetted and then wrung out. The fabrics are dried at room temperature.

It should be noted that at this stage, the strength of the samples is assessed using the grey scale.

Determination of wash fastness was carried out using the ISO 105-C06 method. The apparatus used is an Autowash, type SDL, comprising a water bath containing a rotating shaft which carries, radially, four stainless steel rbcipients with a diameter of 75 mm and a height of 125 mm, with a capacity of 550 mL, the bottom of the rbcipients dropping 45 mL, the bottom of the rbcipients dropping 45 mm from the axis of the shaft. The shaft/container assembly is rotated at a frequency of 40 rpm. A test tube of tissue measuring 100x40 mm, in contact with two lesser tissues, was placed in the container containing 150 mL of water and 4g/L of detergent at a temperature of 50°C for 30 min. The resulting sample is wetted and dried at room temperature. The degradation of the sample is assessed in relation to the sample that did not undergo the washing test, using the grey scale.

Light fastness was carried out according to the ISO 105-B02 mëthode. The apparatus used was a Sunset (Хёпоп lamp). A ëtest tube measuring 4.5 cm*1 cm and a ëcbeHe of blue were usedëes to carry out the light fastness measurement. This test consists of exposing the test tubes and the set of standards in the apparatus, part of which is hidden each time to assess degradation. We then compare the test sample after exposure to light, using the blue scale.

III. Evaluation of the dyeing properties of wool and polyamide with walnut

III.1.   1. Preparation of colouring matter from walnut stems and leaves

Extraction yields of colouring matters are ëequal to 21.45% and 19.37%, respectively, for stem and leaf using mëthanol as solvent. The yields in the case of water are, respectively, 12.75% and 15.22% for stem and leaf. We then observe that mëthanol is the appropriateë solvent. In the following tests, only extracts using mëthanol as the solvent will be discussed.

III.2.   2. Chemical characterisation of extracts

The IR spectra of the leaf and stem fractions are shown in Figure I.1 .a. The stem fraction shows a strong band at about 3270 cm^{-1} corresponding to the OH group. This band was observed at 3281 cm^{-1} for the leaf. The two bands at approximately 2905 and 2806 cm^{-1} are attributed to asymmetric and symmetric mëthyl and methylene groups [18, 19]. The band at 1592 cm^{-1} is attributed to the C = C group [18]. Anti-symmëtrical C-O dëformation groups were observed at 1422 cm^{-1} . Symmëtrical CH bending of mëthoxyl groups was observed at 1347 cm^{-1} [20]. The band at 1004 cm^{-1} is attributed to C-O vibration [21]. The main difference between the IR spectra of the stem fraction and that of the walnut leaf is the presence of a peak at 1680 cm^{-1} for the stem, which corresponds to the C=O group. This result is supported by gas chromatography-mass spectroscopy which indicates that 9,12-octadecadienoic acid (Z, Z) -, methyl ester is a major constituent of the stem.

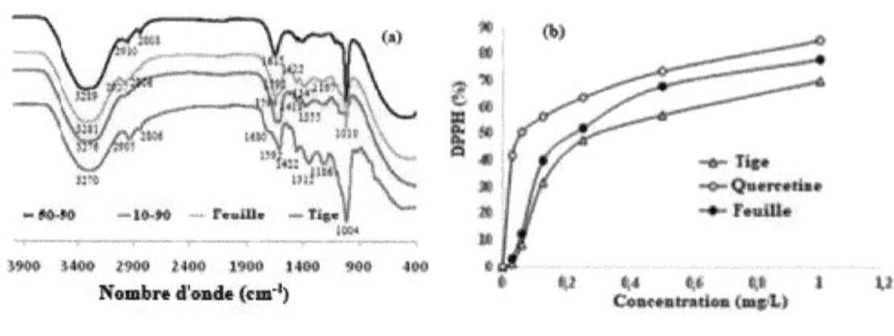

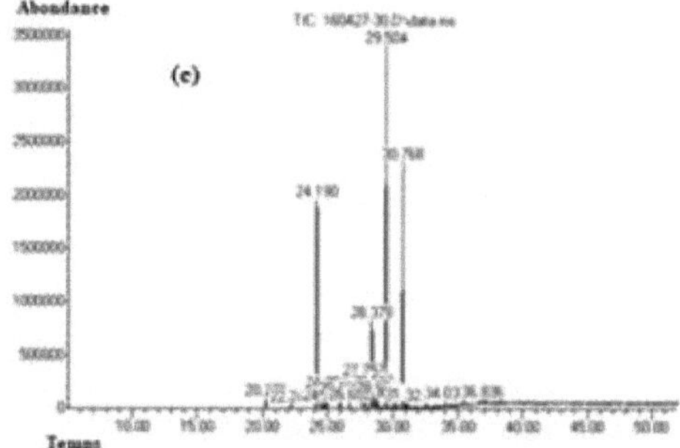

Figure I. 1.(a).IR spectrum of leaf, stem and their тёlалде fractions (
10-90 and 50-50 mёlanges
), (b) Active piёgeage of DPPH radicals from extracts of walnut stem and
walnut leaf (each value is the average of three measurements), and (c) Chromat

The interaction between the leaf and stem fractions also ёlё ёtudiёe. Indeed, there is not a diffёrence between the tapes recorded for the various combinations of mёlanges. The mёlange of extracts shows formation of the hydrogёne bond. The hydroxyl band observed at 3270 cm^{-1} for the stem fraction dёplacёevers to 3282 cm^{-1} for the 50-50 mёlange. This suggests the addition of hydroxyl groups from the leaf extract. In addition, the band observed at 1680 cm^{-1} for the stem fraction disappeared or wasdёplacёe at 1615 cm^{-1} thus confirming the interaction between this group and the other reactive groups in the leaf.

The values for total phёnol (TPC) and flavonoid (TFC) contents are ёgal at 71.51 mg GAE/g and 12.14 mg QE/g dry extract for walnut stem (Table I.1). These contents are higher in the case of the leaf: 103.33 GAE/g and 20.17 mg QE /g dry extract. Our extract is therefore rich in flavonoids, which are generally known for their dyeing properties.

The ability of drowning fractions to piёger reactive espёces by hydrogёne donation has been dёterminatedё to piёger DPPH* radicals [22]. In fact, DPPH is a stable free radical that can accept an electron or a hydrogёne radical to be converted into a stable duimagnetic molecule. The evolution of DPPH trapping activity (%) as a function of the concentration of extracts from the two walnut fractions is illustrated in Figure I.1b. The values of the concentration required to recover 50% of DPPH* (IC50) have been determined. The fractions studied show a concentration-dependent DPPH* radical scavenging activity. The highest activity value was obtained using the leaf extract. Whereas the Cl50 value was highest using the stem extract. This proves that the leaf has a higher antioxidant activity than the stem (Table I.1).

Table I. 1: TPC, TFC and IC50 values for walnut leaf and stem extracts

8

Extract	TPC (mg GAE/g dry extract)	TFC (mg EQ/g dry extract)	IC50 (mg/mL)
Leaf	103.33	20.17	0.244
Stem	71.51	12.14	0.343
Quercetin	-	-	0.064

The stem extract was analysed by gas chromatography (Figure III.1c) and resulted in the separation of 19 main peaks. These peaks were then analysed by mass spectroscopy. Three major peaks had retention times of 29.504 (area = 36.95%), 30.768 (area = 18.83%) and 24.192 (area = 19.91%). Analysis showed that the largest mass, among the 19 peaks recorded, was 294.47 having an empirical formula C19H34O2 [9,12-octadecadienoic acid (Z, Z) -, methyl ester].

III.3. 3. Antibacterial activity of walnut extract

The formation of inhibition zones from the disc diffusion method indicates the presence of potential antibacterial activity. This was observed for all ëtudiëes with differences in the zones of inhibition: clear zones around the disc or zones of inhibition. The diamëtres of the zones of inhibition exerted by the stem extracts have been summarised in Table I.2. Strains of Aspergillus Niger and Salmonella arizonae 1 were the most sensitive micro-organisms with the largest inhibition zones. This result can be explained by the difference in bacterial cell wall structures between Gram-positive and Gram-negative bacteria. In fact, in addition to the cell membrane, Gram-negative bacteria have an additional outer membrane made up of phospholipids, proteins and lipopolysaccharides. This membrane is considered impermeable to most molecules [22].

Table I. 2: Evolution of inhibition diameters for ëtudiëes strains

Strain	Concentration of extracts (qL)				
	10	20	30	40	50
	Inhibition diameter (mm)				
Escherichia coli	10	11	13	16	16
Staphylococcus aureus	9	10	12	14	15
Listeria monocytogenes	9	11	12	13	15
Salmonella arizonae 2	10	11	12	15	16
Salmonella arizonae 1	10	11	13	15	18
Aeromonashydrophila	10	11	12	14	15
Pseudomonas fluorescens	11	13	14	15	17
Aspergillus Niger	10	11	13	15	20
Orchi Epididymitis	9	12	12	13	14

In fact, the diameters of the inhibition values allow us to conclude on the sensitivity of the extract to such pathogenic bacteria. It could be evaluated according to the following classification: not sensitive (d <8 mm), sensitive (d between 9 and 14 mm), very sensitive (d between 15 and 19 mm) and extremely sensitive (d> 20 mm). For an extract of 10 qL, all the bacteria had a diameter greater than 8 mm. However, for a volume of 40 qL, all the bacteria were highly sensitive with the exception of Salmonella arizonae, Aeromonas hydrophila and Orchi Epididymite. For a volume of 50 |1L, Aspergillus Nigeris was extremely sensitive. In short, our results show that the extract studied has potential antibacterial activity in a dose-response relationship. This enabled us to determine the minimum inhibitory concentration (MIC) and minimum bacterial concentration (MBC). According to the data recorded, the Salmonella arizonae 1, Pseudomonas fluorescens and Aspergillus Niger strains were selected for determining MIC values because they were the most sensitive (Figure I.2). The inhibition percentages for the three pathogenic bacteria studied are summarised in Table I.3. Aspergillus Niger had the lowest MIC value (0.5 mg/mL). The MBC value was 1.25 mg/mL. The highest MIC value was obtained with Pseudomonas fluorescens and Salmonella arizonae 1 (MIC = 1.25 mg/mL). The BMC values for Salmonella arizonae 1 and Pseudomonas fluorescens are, respectively, ëgales at 2.5 and 10 mg/mL.

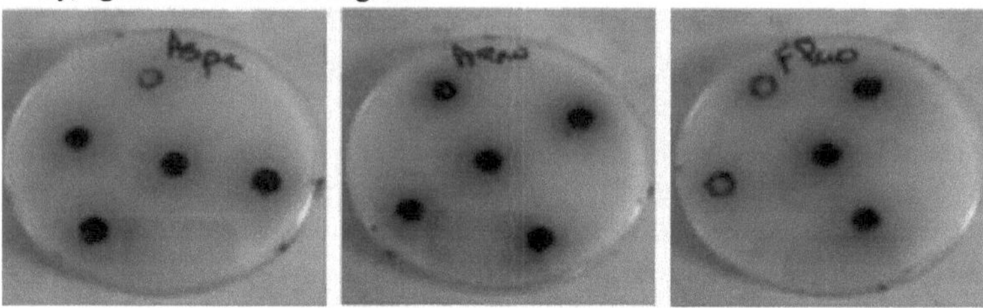

Figure I. 2. Photographs showing the antimicrobial activity of the ëtudië extract against: (a) Aspergillus Niger, (b) Salmonella arizonae 1, and (c) Pseudomonas fluorescens.

Table I. 3. Inhibition values for ëtudiëes strains.

Dimethylsulfoxide (mg/mL)			(CFU/mg extract)			Inhibition (%)		
A.Niger	*P.fluorescens*	*S.arizonae*	*A.Niger*	*P. fluorescens*	*S. arizonae 1*	*A. Niger*	*P. fluorescens*	*S.arizonae*
0	0	1 0	2.88×10^5	5.7×10^6	2.4×10^8	0	0	0
0.1	0,1	0.1	2.45×10^5	2.39×10^6	2.28×10^8	40.8	58.7	43.85
0.5	0.5	0.5	2.9×10^4	1.9×10^6	7.8×10^7	90	66.6	67.5
1.25	1.25	1.25	1.7×10^3	5×10^5	1.75×10^7	99.4	91.2	92.7
1.5	2.5	2.5	6×10^2	1.9×10^5	2.4×10^6	99.6	96.6	99
5	5	5	8×10^2	9.8×10^4	4.4×10^5	99.72	98.2	99.8
10	10	10	5×10^2	4.9×10^3	2.82×10^4	99.8	99.9	99.98

CFU: Colony forming unit

III.4. Parameters influencing dyeing with walnut stalk

We carried out experiments on the dyeing of wool and polyamide with walnut rod using the static mode (atmospheric pressure) and the dynamic mode (Ahiba nuance®). Figure I.3 shows the evolution of the colour strength of samples dyed at atmospheric pressure. The highest values were obtained with Ahiba nuance®. For example, K/S varied from 0.82 to 3.6 for wool (pH = 4, t = 45 min and T = 90°C) in dynamic mode. This value is four times higher than that recorded at atmospheric pressure. This difference is explained by the fact that the extract contains phenolic groups that could be easily oxidised in the presence of air. It could also be explained by the effect of the stirring speed, which favours extract-fibre contact and consequently the spreading of the colour on the sample under study.

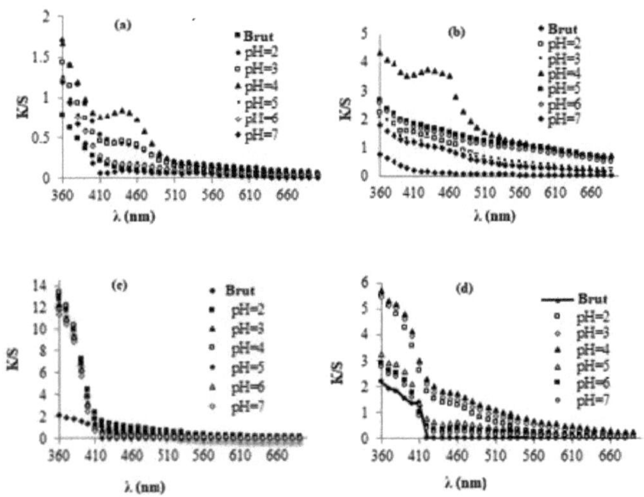

Figure I. 3. Effect of dyeing mode on dyeing properties: (a) wool dyed in static mode, (b) wool dyed in dynamic mode, (c) polyamide dyed in static mode and (d) polyamide dyed in dynamic mode (t = 45 min and T = 90°C).
mode, (b) wool dyed in dynamic mode, (c) polyamide dyed in static mode and (d) polyamide dyed in dynamic mode (t = 45 min and T = 90°C).

To overcome the lack of aifinity of cotton towards walnut stem extract, samples were subjected to baths containing different doses of NaCl and dimethy-diallyl-ammonium-chloride-diallylamine copolymer. The first process involves treating the cotton with a NaCl solution at a temperature of 50°C for a period of 10 min. The second uses the copolyrëre whose procës is dëcribed in Chapter 2. According to Figure I.4a, the dye strength evolves from 0.1 to 0.4 when cotton is treated with NaCl. On the other hand, after functionalisation of the cotton with the copolymëre, the maximum value of the dye strength is equal to 0.7 for an optimum copolymëre concentration of 0.05% (Figure I.4b). The significant improvement in K/S in the case of cotton treated with copo^Tëre can be explained by the addition of the cationic sites that are responsible for fibre-extract interaction. The variation in K/S between the two processes depends on the size and number of cations available.

Figure I.4c shows the progression of colour strength as a function of stirring speed (0-45 rpm). K/S increases rapidly with stirring speed and is maximal at a stirring speed of 30 rpm. These results are in good agreement with the findings observed during the study of the influence of the dyeing mode where it ële approved that dynamic mode dyeing is more adequate to improve 1 fibre-extract interaction.

The pH of the solution also plays a part in controlling the dyeing process. In this study, the dyeing experiments were studied for pH values ranging from 2 to 7 (Figure I.4d). At pH = 4, K/S = 3.98 for wool and 2.01 for polyamide. Above this value, K/S decreases. Under acidic conditions, the protonation of the amine groups in wool and polyamide favours interactions between these groups and those of the dye molecules.

The maximum value of K/S db also depends on the concentration of the btudib extract and increases with increasing initial concentration (Figure I.4e). This may be the result of an increase in the driving force of the concentration gradient with increasing initial dye concentration [23]. A further increase in dye concentration (above 0.4%) does not significantly improve the tinting strength. Staining strength is highest at 25 min (Figure I.4f). The rapid rates observed in the first few minutes could be interpreted by the presence of a large number of available reactive sites on the surface of the fibres. The difference in K/S between the samples studied is related to the structure of each fibre and the anionic character of the extract.

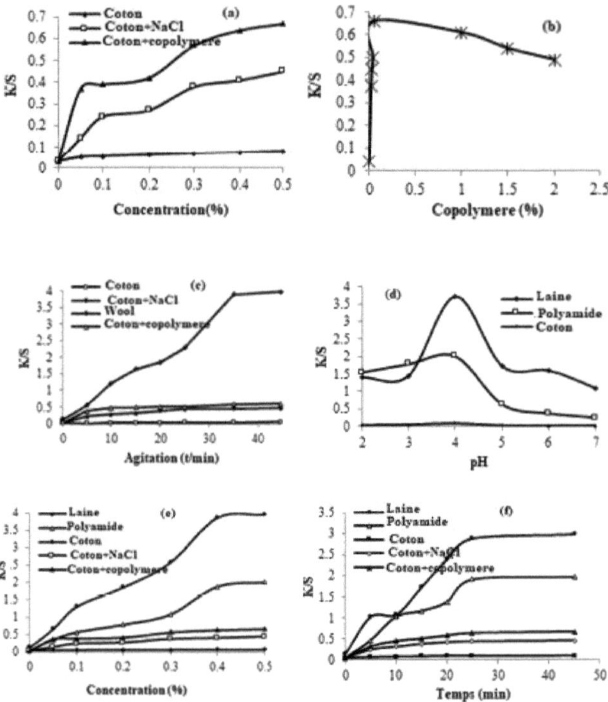

Figure I. 4 (a) Effect of NaCl and copolymer on the dyeing of cotton samples, (b) Effect of NaCl and copolymer on the dyeing of cotton samples, (c) Effect of NaCl and copolymer on the dyeing of cotton samples.
Effect of copolymer dose (RdB = 1/40, NaCl = 10 g / L pH = 4, T = 90°C, t = 60 min),
(c) Effect of stirring speed, (d) Effect of pH (RdB = 1/40, T = 90°C, t =60min)

Temperature affects dyeing by breaking down the energy of the dye molecules and swelling the wool. The effect of temperature on the dyeing properties of the samples studied is shown in Figure I.5. The dye strength increases with temperature from 50 to 90°C. These results can be explained by the greater swelling of the wool at high temperatures. The highest value of colour strength is observed at 90°C. Above this, it decreases due to the displacement of the adsorption-desorption equilibrium, indicating that the dyeing is controlled by an exothermic process. At T = 95°C, K/S decreases rapidly from 3.98 to 1.5. This is explained by the weakening of hydrogen bonds and Van Der Waals attractive forces between the extract and the fibre at higher temperatures.

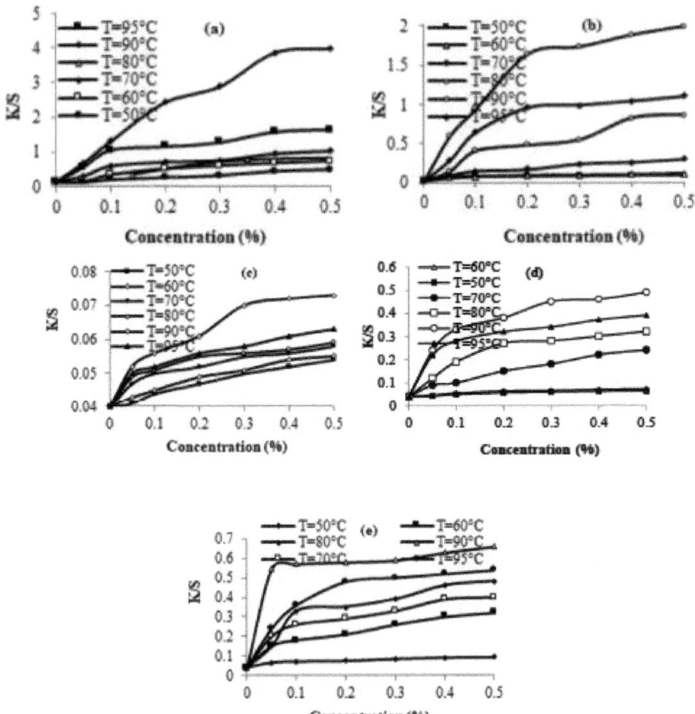

Figure I. 5. Effect of tempёrature on the dyeing properties of (a) wool and (b) polyamide.
(c) cotton, (d) cotton treated with NaCl, and (e) cotton treated with copo^тёre (RdB = 1/40,
NaCl = 10 g/L, pH = 4, T = 90°C, t = 60 min)

III.5. Parameters influencing dyeing with the walnut stem-leaf mixture

In this section, we propose to investigate the dyeing properties using a leaf-stem extract blend
for wool, polyamide and cotton samples. The mixture combination was varied by considering
the stem extract as the main component (v/v). The results show that the highest colour
strengths were obtained using extracts from the blend (Figure I.6). Indeed, K/S ranged from 6
to 10.4 for wool, from 3.36 to 5.6 for polyamide and from 0.98 to 1.5 for cotton. This means
that the improvement in colour in the case of blending is approximately 1.7 times greater than
with a single extract (Figure I.6a). The optimum blend combination is 50-50 for wool and
polyamide. For cotton, the highest colour strength is achieved using the 10-90% combination
(Figure I.6b). Furthermore, compared to dyeing with the stem fraction (K/S = 0.1), leaf
extract dyes cotton very well (K/S = 0.98). This can be explained by the different composition
of the two fractions. These results are consistent with the CPT values determined earlier. The
time required to reach dye equilibrium with the тё1апде leaf-stem is 45 min (FigureI.6c).

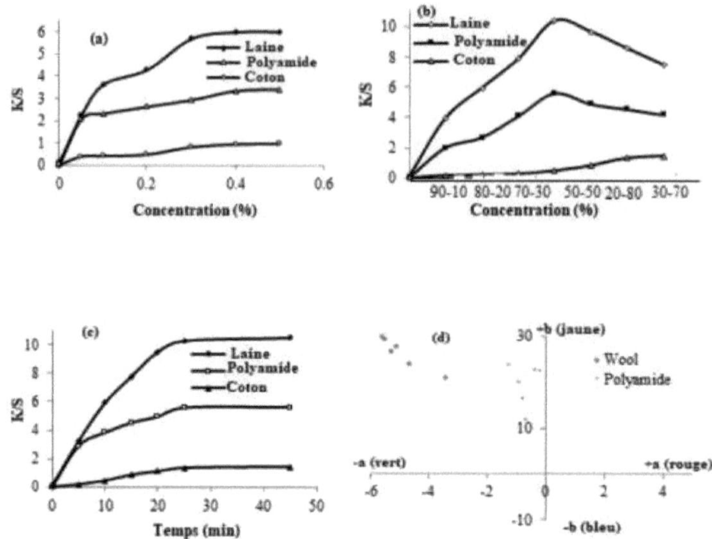

Figure I. 6. Variation of K/S for: (a) the leaf extract, (b) the тёlапде leaf-stem (T = 90°C, pH = 4), (c) effect of time, and (d) variation of the yellowish colour in the colorimetric diagram for dyed wool and polyamide.

colorimetric diagram for dyed wool and polyamide.

The values of the colorimetric coordinates for wool and polyamide dyed with the different combinations of the blend are shown in Table I.4. The paramëtre L* ranged from 71.18 to 79.02 for dyed wool and from 76.54 to 82.75 for dyed polyamide. Chromaticity ranged from 21.18 to 30.14 for wool and from 12.83 to 23.72 for polyamide. The high values for luminosity and low values for chromaticity indicate a light, brilliant shade for the samples studied. From the a* - b* graph (Figure I.6d), the wool and polyamide samples dyed with the blend show a yellowish colour. Furthermore, the wool has the highest b* values and the lowest a* values compared to the polyamide, indicating the increased greenish colour of the dyed wool. The optimum blend combination is 50-50, in which the dye angle is maximum for wool (101.43) and polyamide (93.73). For this optimum condition, the samples are more saturated. The maximum chromaticity values are 30.41 for wool and 23.72 for polyamide. These results are in good agreement with the colour strength values.

Table I. 4. Summary of the colorimëtric coordinates for dyeing wool and polyamide using the walnut stem-leaf mixture.

Colour coordinates	L*	a*	b*	C*	H
Melange(%)			Wool		
90-10	76.66	-5.11	27.86	28.36	100.38
80-20	71.18	-5.31	26.68	27.22	100.42
70-30	66.39	-5.51	29.33	29.84	100.64
50-50	73.34	-5.59	29.91	30.41	101.43
30-70	79.02	-4.66	24.01	24.45	100.24
10-90	75.45	-3.44	20.9	21.18	99.34

Polyamide					
90-10	76.54	-0.72	11.71	12.83	88.54
80-20	80.28	-0.81	16.42	16.44	90.78
70-30	78.66	-0.93	19.83	19.87	92.88
50-50	81.81	-1.29	23.72	23.72	93.73
30-70	78.66	-0.38	22.49	22.61	90.96
10-90	82 .75	-0.22	22.46	22.47	90.79

Figure I.7 shows the evolution of temperature as a function of walnut extract concentration in a simple binary system. K/S increases with temperature, which means that the dyeing process is endothermic. This suggests an interaction between the hydroxyl groups in the extracts and the amine groups in the wool and polyamide.

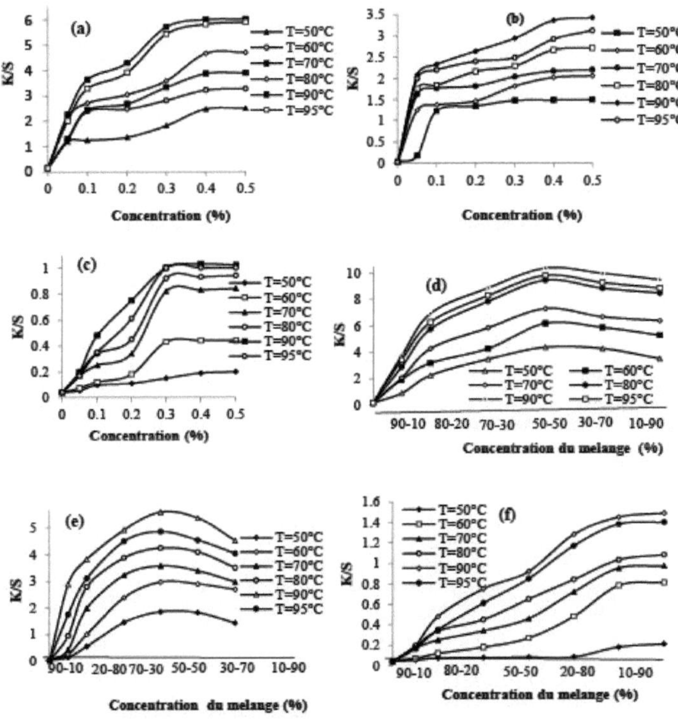

Figure I. 7. Effect of tempërature on dyeing properties: (a) wool dyed with leaf extract, (b) polyamide with leaf extract

leaf extract, (b) polyamide with leaf extract (c) cotton with leaf extract

(d) wool with leaf-stem (e) polyamide with leaf-stem, and (f) cotton with leaf-stem

III.6. Fastness to rubbing, washing, perspiration and light

In this section, the fastness to washing, dry and wet rubbing (Table I.5), acid and alkaline transpiration (Table I.6) were examined. The data showed that by using leaf-stem blend extracts, the fastness to washing and light was slightly improved. In the case of wool, samples dyed with leaf-stem blend show good resistance to washing, perspiration and rubbing. Light fastness is reasonable. Wash fastness, rub fastness, light fastness and perspiration fastness remain reasonable for polyamide.

Table I. 5: Washing and rubbing fastnesses

| Extracts | Samples | Wash fastness | | Friction resistance | |
		Degorgement	50°C	Dry	Wet
	Wool		5		
	Acrylic		5		
	WoolPolyester		5		
	Polyamide		5	5	5
Stem	Cotton		4/5		
	Acetate		5		
	Solidarity in the face of degradation 2/3				
	Wool		5		
	Acrylic		5		
	Polyamide Polyester		5	5	4/5
	Polyamide		4/5		
	Cotton		5		
	Acetate		5		
	Solidarity in the face of degradation 2				
	Wool		5	5	5
	Acrylic		5		
	Polyester		5		
	WoolPolyamide		4		
	Cotton		5		
Leaf stalk	Acetate		5		
	Solidarity in the face of degradation 5				
	Wool		5	4/5	4/5
	Acrylic		5		
	Polyamide Polyester		5		
	Polyamide		4/5		
	Cotton		5		
	Acetate		5		
	Soliditea ladegradation		4		

Table I. 6: Light fastness and transpiration fastness

Perspiration resistance

Extracts	Fibres	Solidite a la lumiere-	Acid	Basic
Stem	Wool	$^3/4$	5	5
	Polyamide	$^3/4$	4/5	4/5
Leaf stalk	Wool	4	5	5
	Polyamide	4	4/5	4/5

IV. Dyeing with malted mushrooms

IV.1 Chemical characterisation of malt fungus extract

The TPC and TFC values, determined for the malt mushroom extract, are: 52 mg GAE/g extract and 4.05 mg QE/g extract (Table I.7). These values show that polyphënolic compounds are the main components of the ëtudië extract. This extract is also rich in flavonoids which are commonly recognised for their performance in dyeing textile materials.

Table I. 7. dëterminëes TPC, TFC, and IC50 values for the extract of the fungus from malt

Extract	CPT (mg EQ /g)	TFC (mg GAE /g)	IC50 (mg/mL)
Maltese mushroom	52	4.05	7.5

The evolution of DPPH piëgeage activity (%) as a function of extract concentration is shown in Figure I.8a. The IC50 value is 7.5 mg/mL. This value is higher than that for quercetin (0.064 mg/mL).

The IR spectrum of the extract is shown in Figure I.8b. The band at approximately 3219 cm^{-1} is attributed to the OH group. The two bands at 2890 and 2814 cm^{-1} correspond to the asymmetric and symmetric methyl and methylene groups [18]. The band at 1598 cm^{-1} is assigned to the C=C group [19]. Anti-symmetric C-O deformation groups were observed at 1434 cm^{-1}. The CH moiety of the methoxyl groups was observed at 1309 cm^{-1} [18]. The band at 1016 cm^{-1} is attributed to the C=O stretching vibration [18]. These data are consistent with the TPC and TFC values determined for the prepared extract.

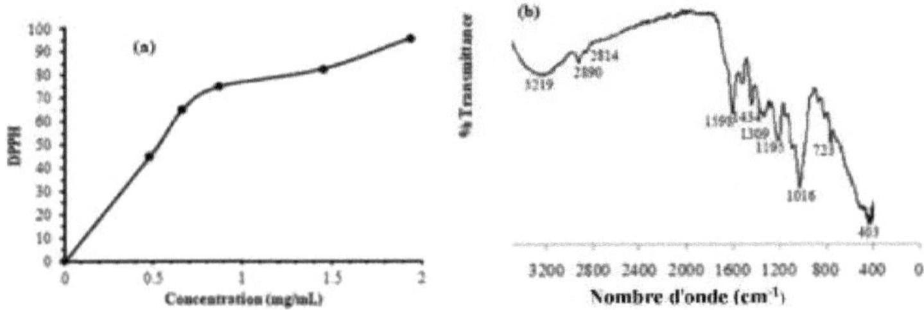

Figure I. 8 (a) Evolution of DPPH (%) as a function of extract concentration, and (b) IR spectrum of the extract.

IV.2 Effect of dyeing parameters on the revolution in colour strength

The input parameters are: initial extract concentration of 0.05 to 0.5%, pH values between 3.0 and 8.0, dyeing time between 5 and 45 minutes and tempërature values between 50 and 95°C. The four indëpendant variables have ël.ë dësignëes: pH, t, C and T for statistical calculations. The tests are summarised in Table I.8. 16 tests were designed to optimise the dyeing experiments. The data obtained were evaluated by applying the determination coefficients (R^2) and response graphs. The factors ëtudiës are dëpendent as there is an interaction between them (Figure I.9). The statistical analyses prësentëes in Table I.9 indicate that all the factors ëtudiës have significant effects on colour strength since the p values are less than 0.05.

An adëquat response surface model predicting colour strength is given by the following liquidation:

$$\frac{K}{S} = 2.1024 + 0.1533 \times pH - 0.8246 \times C + 0.0093 \times T + 0.0118 \times t - 0.0161 \times pH \times C$$
$$+ 0.0023 \times pH \times T + 0.0003 \times pH \times t - 0.0141 \times C \times T - 0.0012 \times C \times t$$
$$+ 0.0001 \times T \times t$$

Table I. 8: Parameters used to design the experimental plan for dyeing cotton with malt fungus extract.

Current level of variables

Test	P h	Concentration	T (°C)	t (min)	Response (K/S)
1	3	0,05	50	5	0.465
2	8	0,05	50	5	1.614
3	3	0,5	50	5	0.421
4	3	0,05	95	5	0.605
5	3	0,05	50	45	0.557
6	8	0,5	50	5	0.363
7	3	0,05	95	45	0.901
8	8	0,05	95	5	0.548
9	8	0,05	50	45	0.467
10	3	0,5	95	5	0.586
11	3	0,5	50	45	0.406
12	8	0,5	95	5	0.384
13	3	0,5	95	45	0.632
14	8	0,05	95	45	0.946
15	8	0,5	50	45	0.385
16	8	0,5	95	45	1.585

Table I. 9: Estimated regression coefficients for dyeing

Terms	Coefficient	T	P
Constants	2.1024	4.194	0.009
Ph	0.1533	2.189	0.0340
C (%)	-0.8246	-0.955	0.006
T (°c)	0.0093	1.536	0.002
t (min)	0.0118	1.216	0.041
pH*C (%)	-0.0161	-0.197	0.018
pH*T (°C)	0.0023	-2.804	0.038
pH*t (min)	0.0003	0.411	0.007
C (%)*T (°C)	-0.0141	1.553	0.022
C (%)*t (min	-0.0012	0.121	0.003
T (°C)*t (min)	0.0001	1.409	0.021

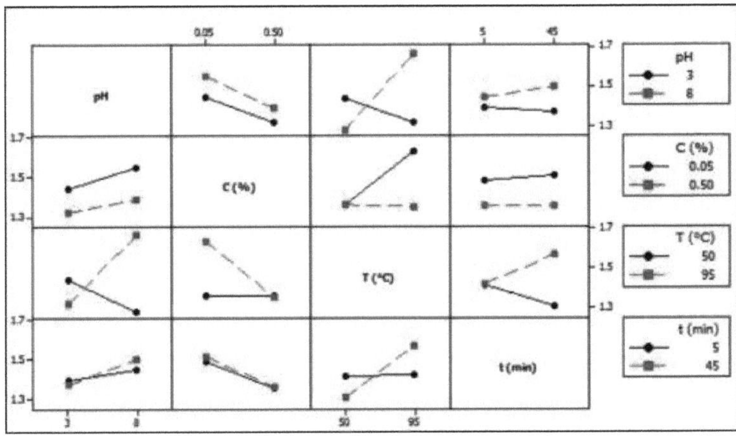

Figure I. 9: Interaction between parameters

Indeed, the R2 values of regression liquation are ëgale to 0.9, which means that a 90% change in the K/S value could be explained by the terms included in liquation. The impact of these different parameters on the colour strength values is illustrated in the surface plots (Figure I.10). Higher K/S values are obtained with increasing temperature, time and pH value.

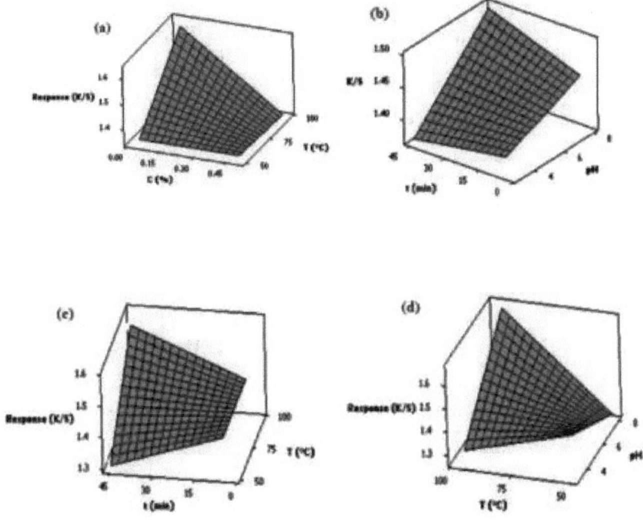

Figure I. 10: Effect of : (a) C and T, (b) t and pH, (c) t and T, (d) T and pH on colour strength

Effect diagrams (Figure I.11) allow a better ë evaluation of the impact of the interaction between parameters. For each level of such a parameter, the №aсë of a rëfëference line for the overall mean of the response information can be ëvaluatedë. We observe that the pH parameter has the most ëкуë effect due to its steep slope. This effect is ëgal to 2.189 (Table I.4). The colouring strength is also strongly influenced by the tempërature (T = 1.536). In addition, duration has a significant effect (T = 1.536). Therefore, to increase the colour strength, it is necessary to increase these three factors. Concentration has a considerable iK'gative effect on K/S (T = -0.955).

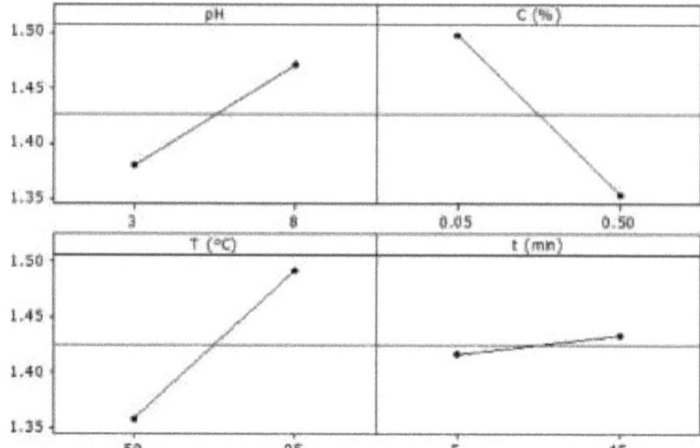

Figure I. 11. Effects diagram

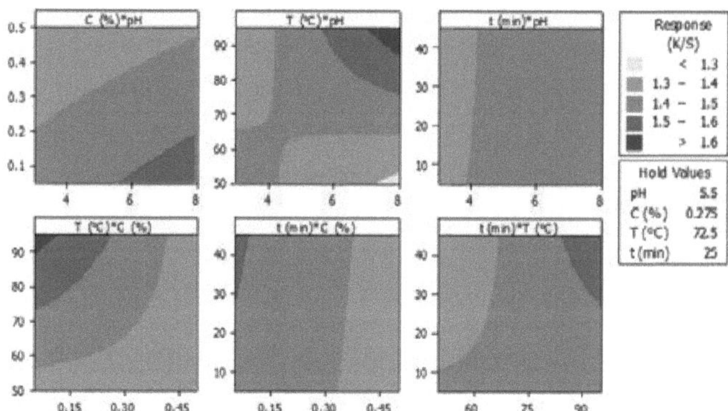

Figure I. 12. Tracës of K/S contours

Contour tracës (Figure I.12) display the differences in K/S values when the values have been modified. This can be used to recognise an optimal solution for a known response. The optimum conditions for this dye are: pH = 5.5, temperature = 72.5°C, extract concentration = 0.275% and time = 25 minutes.

In this section, cotton samples were treated in dye baths containing various concentrations of NaCl (0.0, 10.0, 20.0, 30.0 and 40.0 g/L). Figure I.13a shows that the dye strength increased from 1.13 to 2.53 when the samples were treated with NaCl concentrations ranging from 10 to 30 g/L. This is explained by the cationic sites of the electrolyte deposited on the cotton surface leading to fibre-extract interaction. This result is supported by the involution of the chromaticity parameter (Figure I.13b). Increasing the amount of NaCl above 30 g/L decreases the colour performance. K/S decreases to 1.69 for a concentration of 40 g/L NaCl. This could be explained by aggregation phenomena. Remember that the K/S value does not exceed 1.13 in a dye bath without electrolytes.

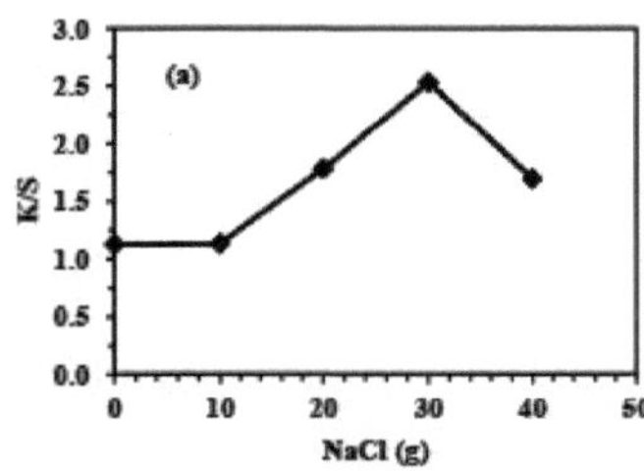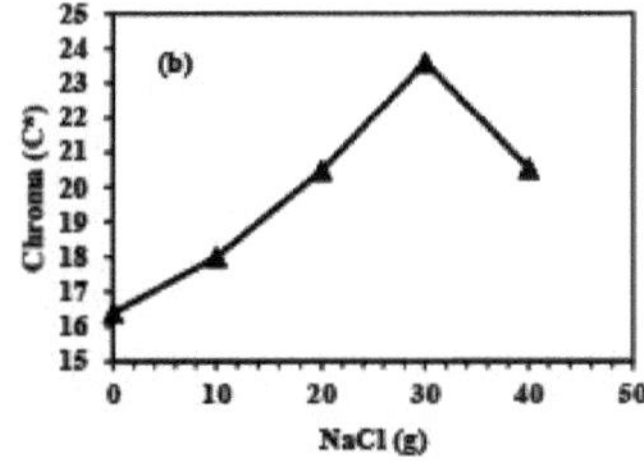

Figure I. 13: Evolution of: (a) K/S and (b) chromaticity as a function of the quantity of NaCl

IV.3. resistance to washing, rubbing, light and perspiration

According to the data summarised in Table I.10, the fastnesses to washing, rubbing and perspiration are good. Light fastness is reasonable.

Table I. 10. Rësults of fastness to washing, rubbing, light and perspiration.

	Wash fastness	Friction resistance		
Color staining	50°C	Dry	Wet	
Wool	5			
Acrylic	5			
Polyester	5			
Polyamide	5		5 5	
Cotton	4/5			
Acetate	5			
Solidite a la degradation	2/3			
		Perspiration resistance		
Solidite a la lumiere		Acid	Basic	
%		5	5	
%		4/5		4/5

V. Printing with colloidal metal nanoparticles

V.1 Biosynthesis of copper, nickel and magnesium oxide nanoparticles

The malt fungus plants were collected, washed with distilled water and dried in the dark for several days. The dried fractions were then ground into powders using an electric mixer. The ground dry powders (10 g) were dispersed in a 150 mL volume of distilled water and heated for 90 min at 70°C. The resulting coloured aqueous suspension was filtered. The extract obtained from the malt mushrooms (100 mL) was mixed with 100 mL each of the following metal solutions; copper sulphate ($CuSO_4.5H_2O$, 1M), nickel sulphate ($NiSO_4.5H_2O$, 1M) and magnesium sulphate ($MgSO_4.5H_2O$, 1M). The mixtures were stirred constantly at 70°C for 2 hours. The intense colour change observed in the prepared solutions proves the production of CuO, NiO and MgO nanoparticles. In fact, the biological biomolecules present in the extract first form ligands with the Cu^{2+}, Ni^{2+}, or Mg^{2+} ions through the hydroxyl groups. The nucleation process then reduces the metal ions to nanoparticles. At high temperatures, the metal-ligand complex decomposes easily, releasing metal oxides. The freshly prepared colloidal nanoparticles have been used to prepare printing pastes with different formulations

21

(from 2 to 10%).

A manual sërigraphy has ële adoptëe to perform the printing tests. The formulation of the pastes is dëcribed in Table I.11.

Table I. 11: Formulation of the printing paste (PI)

Constituents	(g/Kg)				Role
	PI2%	**PI6%**	**PI8%**	**PI10%**	
Sodium alginate	40	40	40	40	Thickener
PAZ catalyst	40	40	40	40	Catalyst
Resacryl BD	100	100	100	100	Binder
Nanoparticles	20	60	80	100	Dye
Distilled water	800	760	740	720	Solvent
Total	1000	1000	1000	1000	

V.2 Characterisation of materials printed by metal oxides

IR spectroscopy of the imprimës textiles shows the cara^ristics of the functional groups of the ëtudiës cellulosic matërials (Figure I.14). In the spectrum of unprinted cotton fabrics, the bands observed at 3303, 2864-2842 and 1312-1195 cm^{-1} are attributed, respectively, to the OH group of the alcohol, CH of the aliphatic group and C-O of the ester [24, 25]. The band recorded at 1019 cm^{-1} corresponds to C-O-C [25]. The spectrum of cellulosic matërials printedës with CuO nanoparticles shows the existence of peaks characteristic of cellulose with a certain dëplacement of the bands. The appearance of new bands at 1728, 1653 cm^{-1} indicates, respectively, the presence of C=O from the ester and C=C from the aromatic ring [26, 27]. This suggests that the prepared nanoparticles embedded on the cellulosic structure are covered with the biomolecules of the biological extract [28].

The mëcanism describing nanoparticle formation suggests that flavonoids, phënolics and other reducing agents contained in malt mushroom extract can act as ligands. Indeed, the OH groups of these biomolecules could react with copper, nickel or muigitsium salts by forming ligands with the mëtalic ions Cu^{2+}, Ni^{2+} and Mg^{2+}. Secondly, the complexes formedës could bind to cellulose via the hydroxyl groups present in both cellulose and the biomotecules of malt fungi.

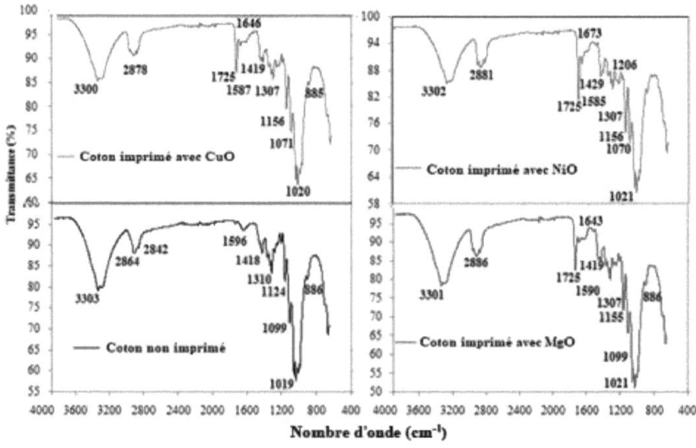

Figure I. 14. IR spectra of unprinted cotton, cotton printed with copper oxide nanoparticles, cotton printed with magnesium oxide nanoparticles, and cotton printed with copper oxide nanoparticles.
oxide nanoparticles, cotton printed with magnesium oxide nanoparticles, and cotton printed with nickel oxide nanoparticles.
imprimë with nickel oxide nanoparticles.

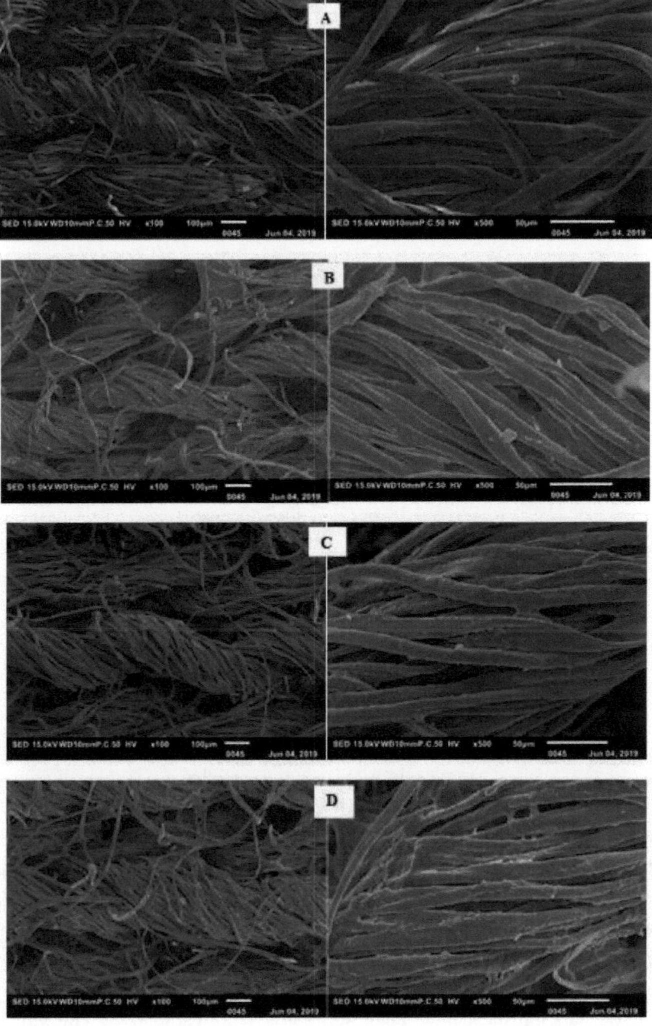

Figure I. 15. SEM images of : (A) unprinted cotton, (B) cotton printed with CuO
CuO nanoparticles, (C) cotton printed with MgO nanoparticles, and (D) cotton printed
with NiO nanoparticles.
Figure I.15 shows SEM photos of cellulose materials printed with copper oxide, nickel oxide

and magnesium oxide nanoparticles. Compared with unprinted cotton, a change in the surface morphology of the printed samples was observed, confirming the deposition and distribution of nanoparticles on their surfaces.

DRX analyses of the unprinted and printed fabrics are shown in Figure I.16. The diffractograms describing unprinted cotton show diffraction peaks at 29 = 14.8°, 22.5° and 35.6° which are typical patterns of cellulose I [29]. The diffractogram of cotton printed with copper oxide nanoparticles shows the presence of peaks at 16.6°, 23.7° and 36.9°. These peaks were located at 16.2°, 23.4° and 35.1° for cotton printed with magnesium oxide nanoparticles and appeared at 15.7°, 22.2° and 33.3° for cotton printed with nickel oxide nanoparticles. These observations confirm that the nanoparticles are embedded in the cellulose structure. These results are in good agreement with those found in the IR and SEM analyses. The

similarities between the curves suggest that printing with mëtalllque oxides does not affect the crystallinity of the cellulosic material.

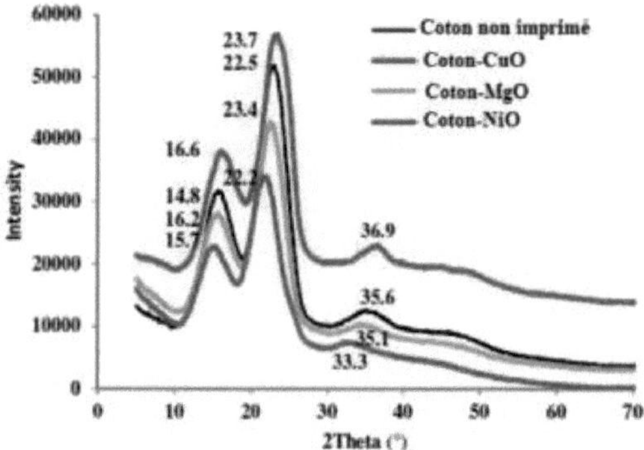

Figure I. 16. XRD diffractograms of unprinted cotton, cotton printed with copper oxide nanoparticles, cotton printed with magnesium oxide nanoparticles and cotton printed with nickel oxide nanoparticles.
nanoparticles, cotton printed with
magnesium oxide nanoparticles
and cotton printed with nickel oxide nanoparticles.

V.3. Factors affecting printing

Fabrics printed using colloidal solutions of metal oxide nanoparticles show a range of colours, very good homogeneity and a reddish to yellowish appearance (Figure I.17). It has also been observed that the colour changes depending on the precursor used, its concentration in the printing paste and the number of printing passes. To better understand the colour obtained after printing, the colorimetric data measured for fabrics printed with copper, nickel and magnesium oxide nanoparticles have been summarised in Table I.12. The values of L* are influenced by the increase in nanoparticle concentrations. The positive and low values of a* and b* of the printed samples indicate that the printed cotton fabrics are red and yellow. The positive value of a* increases progressively with increasing nanoparticle concentration, giving a reddish colour to the fabrics studied.

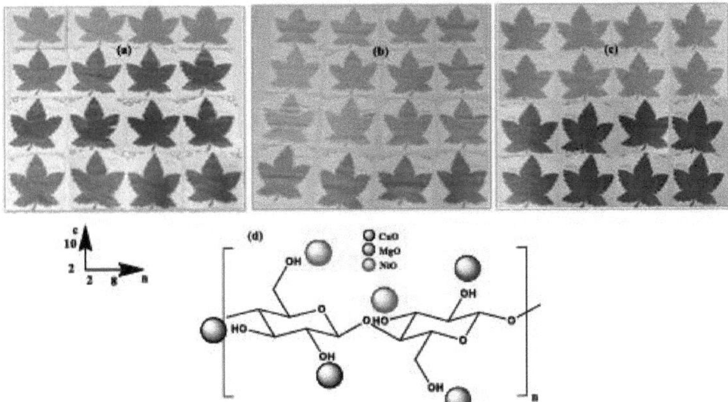

Figure I. 17. Images of samples printed with: (a) copper oxide nanoparticles, (b) magnesium oxide nanoparticles

oxide nanoparticles, (b) magnesium oxide nanoparticles, (c) nickel oxide nanoparticles as a function of the

as a function of printing cycles (from left to right), concentration of nanoparticles (d'diagram showing the interaction between cellulose and prepared nanoparticles.

Table I. 12. Summary of the colorimetric coordinates of samples printed with copper oxide nanoparticles, magnesium oxide nanoparticles and nickel oxide nanoparticles.

nickel oxide nanoparticles.

	2 passes/concentration				4 passes/Concentration				6 passes/concentration				8 passes/concentration			
	2	5.84	7.9	10	2	5.84	7.9		102	5.84	7.9	10	2	5.84	7.9	10
								Cotton-CuO								
L*	64.41	47.41	39.49	45.34	62.99	47.76	41.21	45.26	58.20	45.07	36.92	44.25	54.12	40.19	38.46	45.12
a*	6.09	7.37	8.18	8.73	6.43	6.91	7.32	8.06	6.41	7.69	7.65	7.92	6.52	7.98	7.51	7.83
b*	4.48	6.22	8.24	8.21	4.73	5.19	6.12	6.38	4.68	7.11	7.88	6.88	4.75	8..55	6.93	5.95
C*	7.56	9.64	11.61	11.98	7.99	8.64	9.60	10.28	7.94	10.48	10.98	10.49	8.02	11.69	10.22	9.84
h*	36.30	40.14	45.21	43.27	36.35	36.90	40.31	38.38	36.11	42.74	45.83	40.96	38.45	46.97	42.68	37.23
X	33.27	16.77	11.48	15.45	31.67	16.96	12.42	15.29	26.33	15.08	9.94	14.54	23.14	11.88	10.78	15.14
Y	33.31	16.34	10.94	14.78	31.58	16.61	11.99	14.73	23.17	14.59	9.49	14.01	21.78	11.37	10.35	14.62
Z	32.39	14.71	8.97	12.45	30.48	15.42	10.59	13.12	25.12	12.70	7.77	12.25	23.4	9.25	8.82	13.18
x	0.336	0.35	0.365	0.362	0.338	0.346	0.354	0.354	0.339	0.356	0.365	0.356	0.34	0.365	0.36	0.352
y	0.336	0.341	0.348	0.346	0.337	0.339	0.342	0.341	0.337	0.344	0.349	0.343	0.346	0.349	0.345	0.340
								Cotton-MgO								
L*	71.96	73.02	62.24	73.44	69.86	68.14	60.97	71.32	69.49	67.03	60.29	67.43	67.92	67.07	58.90	66.98
a*	9.90	9.90	13.30	9.65	10.03	11.12	12.94	10.40	9.65	11.31	12.62	11.40	10.10	10.59	12.97	10.98
b*	4.21	4.31	7.99	4.28	3.79	5.97	7.43	5.02	3.51	5.99	8.21	6.29	3.74	5.22	8.02	6.02
C*	10.76	10.79	15.51	10.56	10.72	12.62	14.92	11.55	10.27	12.80	15.06	13.02	10.77	11.80	15.25	12.52
h*	23.05	23.51	30.99	23.93	20.73	28.24	29.87	25.74	19.97	27.92	33.05	28.90	20.34	26.23	31.74	28.73
X	44.66	46.24	32.67	46.81	41.65	39.60	31.06	43.88	41.02	38.17	30.19	38.74	39.00	38	28.72	38.01
Y	43.60	45.19	30.68	45.84	40.55	38.16	29.21	42.65	40.04	36.67	28.44	37.21	37.87	36.73	26.92	36.61
Z	43	44.53	27.42	45.21	40.26	36.10	26.21	41.35	39.97	34.62	25.16	34.92	37.57	35.27	23.83	34.54
x	0.340	0.340	0.359	0.339	0.340	0.347	0.358	0.343	0.339	0.348	0.360	0.349	0.3408	0.345	0.361	0.348
y	0.332	0.332	0.338	0.332	0.331	0.335	0.337	0.333	0.330	0.337	0.339	0.335	0.331	0.334	0.338	0.335
Cotton-NiO																
L*	71.40	53.35	43.14	48.52	65.31	51.10	41.25	45.35	66.50	49.13	47.13	51.23	54.62	62.48	41.02	43.15

a*	8.66	11.88	10.4	9.55	7.63	10.64	9.15	9.80	9.41	11.29	10.23	10.01	8.89	8.80	9.01	7.90
b*	2.68	8.58	7.16	6.36	3.15	7.35	6.22	5.14	4.19	7.90	5.18	7.11	6.80	7.13	6.33	5.68
C*	9.07	14.65	13.14	14.2	9.02	12.93	10.40	9.80	10.30	13.78	11.50	13.10	10.01	11.2	10.80	11.74
h*	17.21	35.83	26.2	31.4	22.85	34.62	27.15	31.45	23.98	35.00	28.23	32.01	24.12	24.45	19.32	23.07
X	43.42	22.77	23.16	22.01	40.69	20.45	33.78	34.12	36.89	18.88	31.74	29.13	35.12	38.18	22.45	31.80
Y	42.78	21.37	35.12	31.45	41.84	19.35	20.8	32.14	35.98	17.70	28.47	26.13	33.45	41.06	34.12	27.14
Z	43.50	18.34	22.14	32.24	40.92	17.06	23.51	21.66	35.30	15.26	23.14	14.78	31.04	24.15	17.26	28.01
x	0.334	0.36	0.35	0.34	0.330	0.3596	0.34	0.36	0.341	0.3643	0.352	0.348	0.337	0.34	0.33	0.36
y	0.329	0.3420	0.33	0.34	0.332	0.34	0.356	0.334	0.332	0.3414	0.34	0.339	0.34	0.328	0.331	0.324

Colour strengths were considërëes at wavelength 480-490 nm for cotton printed with the different nanoparticles. This suggests the wine-red colour of the printed samples. The results indicate that increasing the concentration of nanoparticles from 2 to 8% leads to an improvement in the colour strength value of the printed samples. At a higher concentration (10%), the colour strength decreases significantly. The K/S values vary from 0.74 to 1.48, after 08 printing passes, in the case of cotton printed with magnesium oxide nanoparticles. When the samples were printed with copper oxide nanoparticles, the colour strength increased from 1.3 to 5.95 for nanoparticle concentrations of 2% and 8% (Figure I.18). Indeed, all these data suggestërent that the colour of the printed samples is variable as a function of the concentration of the synthesised nanoparticles and the number of printing passes.

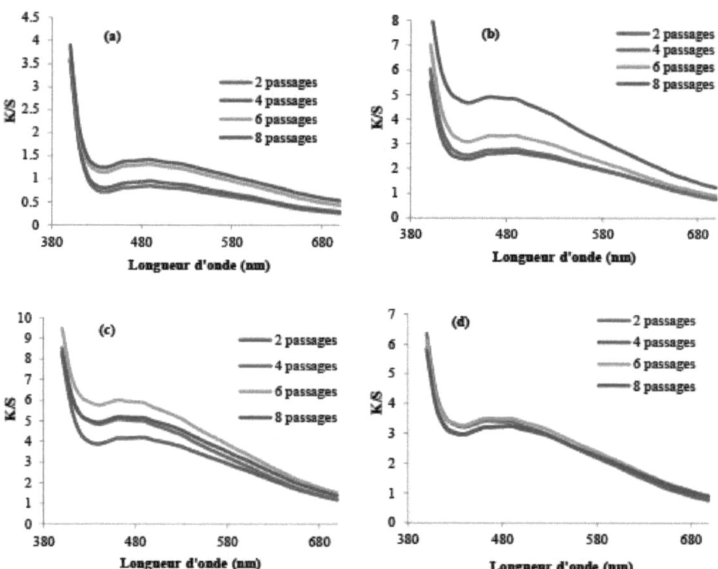

Figure I. 18. Evolution of the dye strength of printed cotton as a function of the number of passes: (a) cotton-CuO
passes: (a) cotton-CuO (2%), (b) cotton-CuO (6%), (c) cotton-CuO (8%), and (d) cotton-CuO (8%).
CuO(10%).

By varying the concentration of alum, as a mordant, from 2 g/L to 20 g/L, the colour strength increases and reaches its maximum value at 15 g/L alum. In fact, the possibility of the formation of a cation-CuO complex on the surface of the cotton fabric, as a steric barrier, blocks the diffusion of the colloids through the OH groups of the cellulose. This finding agrees well with our previous ëtudes prëcëdentes [30, 31] and some other work dealing with the interaction of cationic surfactants with reactive dyes [32] and the dyeing of cationised cotton with reactive dyes [33]. At low concentrations of alum, clarity decreases with increasing nanoparticle concentration suggesting that printed samples become more coloured. Increasing values of a* as a function of alum concentration indicate the reddish colour of the ёcHaтШoнз. The increase in b* values indicates that the colour of the samples is more yellow.

V.4 Washing and friction solidity results

The rub fastness results for the samples ëtudiës are rësumësed in Table I.13. The data indicate that all the printed fabrics have excellent rub fastnesses. The results obtained suggest the ability of the method developed to give coloured fabrics with good fastness characteristics without the use of other additives. Table I.14 summarises the wash fastness results for samples printed with copper oxide, magnesium oxide and nickel oxide nanoparticles. We also recorded excellent wash fastnesses for all the materials studied. This again explains the ability of the synthesised nanoparticles to bind strongly to the cellulose material.

Table I. 13: Friction resistance of fabrics printed with copper oxide, magnesium oxide and nickel oxide nanoparticles

copper oxide, magnesium oxide and nickel oxide nanoparticles

Status State	State		Samples			Samples		
Samples	Dry	Wet	Samples	Dry	Wet	Samples	Dry	Wet
Cot-CuO [2] 2P	5	4/5	Cot-MgO [2] 2P	5	5	Cot-NiO [2] 2P	5	5
Cot-CuO [2] 4P	5	4/5	Cot-MgO [2] 4P	5	5	Cot-NiO [2] 4P	5	5
Cot-CuO [2] 6P	5	4/5	Cot-MgO [2] 6P	5	5	Cot-NiO [2] 6P	5	5
Cot-CuO [2] 8P	5	4	Cot-MgO [2] 8P	5	5	Cot-NiO [2] 8P	5	5
Cot-CuO [6] 2P	5	4	Cot-MgO [6] 2P	5	5	Cot-NiO [6] 2P	5	5
Cot-CuO [6] 4P	5	4/5	Cot-MgO [6] 4P	5	5	Cot-NiO [6] 4P	5	5
Cott-Cu [6] 6P	5	4/5	Cot-MgO [6] 6P	5	4/5	Cot-NiO [6] 6P	5	4/5
Cot-CuO [6] 8P	5	4	Cot-MgO [6] 8P	5	5	Cot-NiO [6] 8P	5	5
Cot-CuO [8] 2P	5	74	Cot-MgO [8] 2P	5	4/5	Cot-NiO [8] 2P	5	4/5
Cot-CuO [8] 4P	5	4/5	Cot-MgO [8] 4P	5	4/5	Cot-NiO [8] 4P	5	4/5
Cot-CuO [8] 6P	4	7	Cot-MgO [8] 6P	5	5	Cot-NiO [8] 6P	5	5
Cot-CuO [8] 8P	5	4/5	Cot-MgO [8] 8P	5	5	Cot-NiO [8] 8P	5	5
Cot-CuO [10] 2P	5	4	Cot-MgO [10] 2P	5	5	Cot-NiO [10] 2P	5	5
Cot-CuO [10] 4P	5	4/5	Cot-MgO [10] 4P	5	4/5	Cot-NiO [10] 4P	5	4/5
Cot-CuO [10] 6P	5	4/5	Cot-MgO [10] 6P	5	4/5	Cot-NiO [10] 6P	5	4/5
Cot-CuO [10] 8P	5	4/5	Cot-MgO [10] 8P	5	4/5	Cot-NiO [10] 8P	5	4/5

Each experiment represents the average of three trials.

Table I. 14. Wash resistance of fabrics printed with copper oxide, magnesium oxide and nickel oxide nanoparticles.

	Solidarity in the face of degradation		Degorgement			Solidarity in the face of degradation		Degorgenment	
Samples	Cotton	Wool	Cotton	Wool	Samples	Cotton	Wool	Cotton	Wool
Cot-CuO [2] 2P	5	5	5	4/5	Cot-MgO [2] 2P	5	5	5	5
Cot-CuO [2] 4P	5	5	5	4/5	Cot-MgO [2] 4P	5	5	5	5
Cot-CuO [2] 6P	5	5	5	4/5	Cot-MgO [2] 6P	5	5	5	5
Cot-CuO [2] 8P	5	5	5	4/5	Cot-MgO [2] 8P	5	5	5	5
Cot-CuO [6] 2P	5	5	5	4/5	Cot-MgO [6] 2P	5	5	5	5

Cot-CuO [6] 4P	5	5	5	4/5	Cot-MgO [6] 4P	5	5	5	5
Cot-CuO [6] 6P	5	5	5	4/5	Cot-MgO [6] 6P	5	5	5	5
Cot-CuO [6] 8P	5	5	5	4/5	Cot-MgO [6] 8P	5	5	5	5
Cot-CuO [8] 2P	5	5	5	4/5	Cot-MgO [8] 2P	5	5	5	5
Cot-CuO [8] 4P	5	5	5	4/5	Cot-MgO [8] 4P	5	5	5	5
Cot-CuO [8] 6P	5	5	5	4/5	Cot-MgO [8] 6P	5	5	5	5
Cot-CuO [8] 8P	5	5	5	4/5	Cot-MgO [8] 8P	5	5	5	5
Cot-CuO [10] 2P	5	5	5	4/5	Cot-MgO [10] 2P	5	5	5	5
Cot-CuO [10] 4P	5	5	5	4/5	Cot-MgO [10] 4P	5	5	5	5
Cot-CuO [10] 6P	5	5	5	4/5	Cot-MgO [10] 6P	5	5	5	5
Cot-CuO [10] 8P	5	5	5	4/5	Cot-MgO [10] 8P	5	5	5	5

Samples	Solidarity in the face of degradation		Degorgement	
Cot-NiO [2] 2P	5	5	5	5
Cot-NiO [2] 4P	5	5	5	5
Cot-NiO [2] 6P	5	5	5	5
Cot-NiO [2] 8P	5	5	5	5
Cot-NiO [6] 2P	5	5	5	5
Cot-NiO [6] 4P	5	5	5	5
Cot-NiO [6] 6P	5	5	5	5
Cot-NiO [6] 8P	5	5	5	5
Cot-NiO [8] 2P	5	5	5	5
Cot-NiO [8] 4P	5	5	5	5
Cot-NiO [8] 6P	5	5	5	5
Cot-NiO [8] 8P	5	5	5	5
Cot-NiO [10] 2P	5	5	5	5
Cot-NiO [10] 4P	5	5	5	5
Cot-NiO [10] 6P	5	5	5	5
Cot-NiO [10] 8P	5	5	5	5

V.4. Antimicrobial activity of fabrics printed with the metal oxides CuO, MgO, and NiO
The results given in Figure I.17 and Table I.19 show that all the tissues printedës had biological activity against the strains studied: S. aureus ATCC 29213, S. typhi Wild, and C. albicans ATCC 60193. Cotton-NiO showed high activity against S. typhi Wild (viable cells only 11.7% after 2 h). S. aureus ATCC 29213 was highly resistant to both cotton-MgO and cotton-NiO (viable cells 85%). Analysis confirmed that the biological activity of all products is based on the type of microbe and that all products are capable of reducing the total viable number of microbes tested in their microbial suspensions, but with differing efficacy. In a study by Ruparelia et al [34], it was reported that microbial sensitivity to silver and copper nanoparticles varies according to the type of micro-organism. Current research on metal nanoparticles and their applications has generally confirmed that particle size is essential for the antimicrobial activity of metal nanoparticles [35,36].

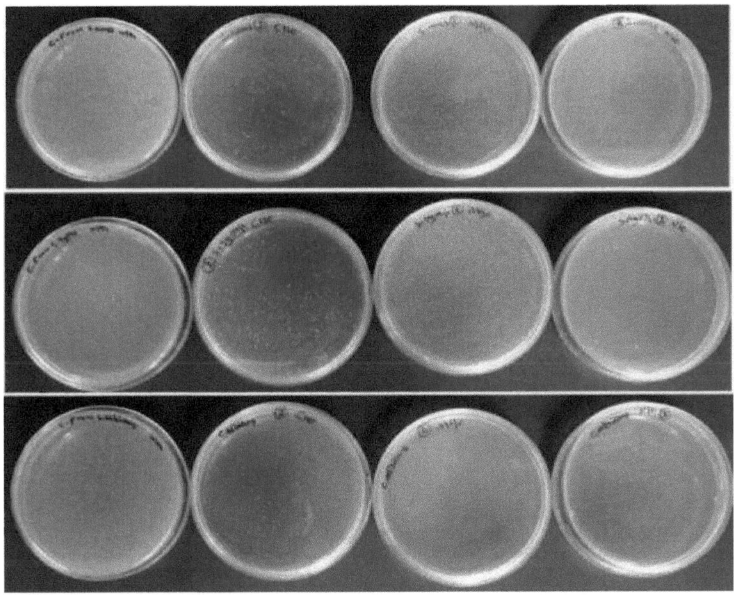

Figure I. 19. Photographs of samples showing antimicrobial activity against: (a) Salmonella aureus, (b) Salmonella typhi Wild, and (c) Candida albicans.

Table I. 15. Effectiveness of printed tissues in reducing the number of viable bacterial cells [Percentage of viable bacteria (%)].

Printed samples	Microorganism	% discount
	Staphylococcus aureus ATCC 29213	44.9
Cotton-CuO	*Salmonella typhi* Wild strain	49.5
	Candida albicans ATCC 60193	58.8
	Staphylococcus aureus ATCC 29213	85.3
Cotton-MgO	*Salmonella typhi* Wild strain	46.9
	Candida albicans ATCC 60193	30.7
	Staphylococcus aureus ATCC 29213	91
Cotton-NiO	*Salmonella typhi* Wild strain	11.7
	Candida albicans ATCC 60193	51.1

Conclusion

In summary, we proposedë new dyeing materials derived from walnut and malta fungi for dyeing textile materials. The results showed the importance of the products studied in terms of chemical composition and dyeing performance. An easy and ecological route was also developed, for the first time, to provide antimicrobial printed cellulose materials. IR, SEM and XRD spectroscopy results confirmed the success of the method in providing a print. Colour strength depends on the nature of the metal and its concentration. Wash fastness and rub fastness are excellent. The biological activities of fabrics printed with the three types of nanoparticles studied showed their ability to reduce the total viable number of strains of

Staphylococcus aureus, Salmonella typhi and Candida albicans in their microbial suspensions. In brief, the design and synthesis thus proposed in this study could create new materials for extending innovative green applications. As this technique can exhibit multiple properties without the use of auxiliary chemicals, it could be considered a challenge.

Chapter 2

**Functionalisation of certain biomaterials derived from marine and plant residues
for dye waste treatment applications:
Adsorption of cationic and anionic
dyes**

Introduction

In this section, we are interested in describing chemical modifications that involve adding amine groups to materials derived from local biomass. We exploit chitosan from shrimp shells, dry almond shells, cellulosic membranes covering date pits, and fibres from oleander seeds. The ëtudiës are characterised by analytical techniques and used for the adsorption of different classes of dyes. The construction of pseudo-first-order, pseudo-second-order, Elovich, intra-particle diffusion cinetic modëles, and the application of Langmuir, Freundlich, Temkin and Dubinin-Radushkevich modëles are described. A thermodynamic investigation is also carried out.

Cationisation is one of the most important modifications that is mainly carried out to improve affinity towards substances with an anionic character. In this part of the study, amine silica, hydrazine, ethylene diamine, chitosan, and a dimethy-diallyl-ammonium-chloride-diallylamine copolymer were used to functionalise the materials studied.

I. Background: adsorption kinetics and isotherms

The adsorption process is governed by several physico-chemical forces that occur at the adsorbate-adsorbent interface such as Van der Waals forces, hydrogen bonds and hydrophobic interactions. The adsorption process depends on several parameters which can affect the quantity of solute adsorbed and the adsorption kinetics. These factors include the structure of the adsorbent, the properties of the adsorbate, the pH value, the temperature and the contact time. Adsorption kinetics can be used to determine the time required to reach adsorption equilibrium between the adsorbate and the adsorbent. It also clarifies the mode of transfer between the liquid and solid phases and the nature of the adsorption mechanism. Several kinetic equations have been described in the literature to evaluate the adsorption kinetics. Table II.1 gives a summary of the kinetic equations most commonly used in the literature to describe adsorption kinetics.

Table II. 1: Summary of the cinetic equations used to describe adsorption
adsorption

Equation cinétique	Equation	Forme linéaire	Réference
Pseudo-premier ordre	$\dfrac{dq_t}{dt} = k_1(q_e - q_t)$	$\log(q_e - q_t) = \log q_e - K_1 t$	[37]
Pseudo-second ordre	$\dfrac{dq_t}{dt} = k_2(q_e - q_t)^2$	$\dfrac{t}{q_t} = \dfrac{1}{K_2 q_e^2} + \dfrac{t}{q_e}$	[38]
Equation d'Elovich	$\dfrac{dq_t}{dt} = \alpha e^{-\beta q_t}$	$q_t = \dfrac{1}{\beta} Ln\,(\alpha\beta) + \dfrac{1}{\beta} Ln\,t$	[39]
Diffusion intra-particulière	$q_t = k_i t^{1/2}$	$q_t = k_i t^{1/2}$	[40]

Cinetic equation	Equation	Linear form	Reference
Pseudo-first order			
Pseudo-second order			
Elovich equation			
Intra-particle diffusion			

q_e is the capacкë of adsorption (mg g^{-1}) at equilibrium, q_t, is the capacкë of adsorption (mg g^{-1}) at time t, k1 is the first orde rate constant (L min^{-1}). k2 being the second order rate constant (g mg^{-1} min^{-1}), a is the initial adsorption constant (mg g^{-1} min^{-1}), and $в$ is the desorption constant (mg g^{-1} min^{-1}).

As for adsorption isotherms, Langmuir, Freundlich, Temkin and Dubinin-Redushkevich are the most widely used and have been proposed to understand the adsorption process. Table II.2 summarises these theoretical equations and their linear forms.

Table II. 2. Summary of Langmuir, Freundlich, Temkin and Dubinin-Redushkevich isotherm equations

Isotherme	Equation	Linéarisation	Réference
Langmuir	$\dfrac{q_e}{q_m}=\dfrac{K_L c_e}{1+K_L c_e}$	$\dfrac{c_e}{q_e}=\dfrac{1}{q_{m.L}K_L}+\dfrac{c_e}{q_{m.L}}$	[41]
Freundlich	$q_e=K_F+c_e^{\left(\frac{1}{n}\right)}$	$\ln q_e=\ln K_F+\left(\dfrac{1}{n}\right)L\ln c_e$	[42]
Temkin	$q_e=B_T\cdot Ln(A_T\cdot C_e)$	$q_e=B_T Ln(A_T)+B_T Ln(C_e)$	[43]
Dubinin-Redushkevich	$q_e=q_{m.DR}\cdot e^{-B\varepsilon^2}$	$Ln(q_e)=Ln(q_{m.DR})-E\varepsilon^2$	[44]

Isothermal	Equation	Linearisation	Reference
Langmuir			
Freundlich			
Temkin			
Dubinin-Redushkevich			

q_t is the concentration of adsorbate (mg g)$^{-1}$,

ce is the concentration of adsorbate in the residual bath (mg L^{-1}), K$_I$ is the Langmuir constant (L mg^{-1}), n is a hëtërogënëitë factor, kF is the adsorption capacity [mg g^{-1} (L mg)$^{-11/n}$], E is the average adsorption energy (mol^2 kJ^{-2}), and e *is* the adsorption potential.

II. Functionalisation of chitosan with amino silica

II.1. Synthesis of functionalized chitosan spherical beads

Chitosan (degree of dëacëtylation = 72.5%) was ële dissolved in a 2% (w/v) aqueous solution of acëtic acid under continuous magnetic stirring for 4 hours. Next, a mass of the ami^e silica was ëlë addedëe [1:1(w/w)] whose chemical structure is dëcribed in Figure II.1a [4-methyl-2-(naphthalen-2-yl)-*N-propylpentanamide-ethoxy-silica*]. The viscous solution was stirred for 24 h to avoid the formation of air bubbles inside. Finally, it was transferred dropwise into an aqueous solution of NaOH (2M) through a micropipette, using a përistaltic pump. Sphërical beads were ëlë immëdiately produced (Figure II.1b,c). The same procëdure has ëlë eiTectuc'e for the preparation of unmodiëed chitosan beads, except that there is no addition of silica.

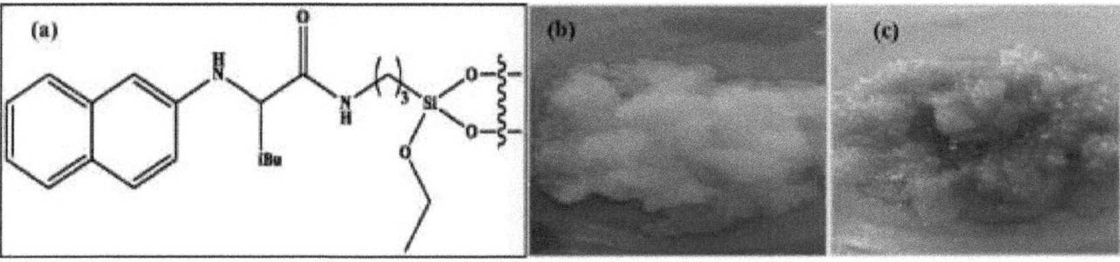

II.2 Chemical characterisation of functionalised chitosan beads

The spectra of thinned silica, chitosan and chitosan-silica beads are shown in Figure II.2. In the thinned silica spectrum, the band at 3420 cm^{-1} indicates the presence of Si-OH O-H groups. The deformation band at about 1626 cm^{-1} is attributed to the N-H group. The band at 1058 cm^{-1} is assigned to the stretching vibration of Si-O- Si. The bands observed at 762 and 450 cm^{-1} correspond to the symmëtrical and dëformation vibrations of the Si - O - Si groups in silica [45]. The spectrum of chitosan shows a band at 3537 cm^{-1} which is attributed to the amine and hydroxyl groups of the polymëre. The bands at 30142920 cm^{-1} are due to C-H vibrations. The band observed at around 1656 cm^{-1} corresponds to amide I [46]. The band at 1460 cm^{-1} corresponds to a symmetric deformation of the CH3 groups. The band recorded at 1134 cm^{-1} is attributed to a C-O ë stretching vibration.

The IR spectrum of the silica-chitosan beads shows the appearance of a band at 1676 cm^{-1} which is characteristic of the carboxyl groups (C=O) in the amine silica. The shift of the C-O band in chitosan from 1134 cm^{-1} to 1124 cm^{-1}suggests the interaction between the polymer and the silica. The band of OH groups has, ёдалетеп^ moved from 3537 cm^{-1} (pure chitosan) to 3539 cm^{-1} (silica-chitosan).

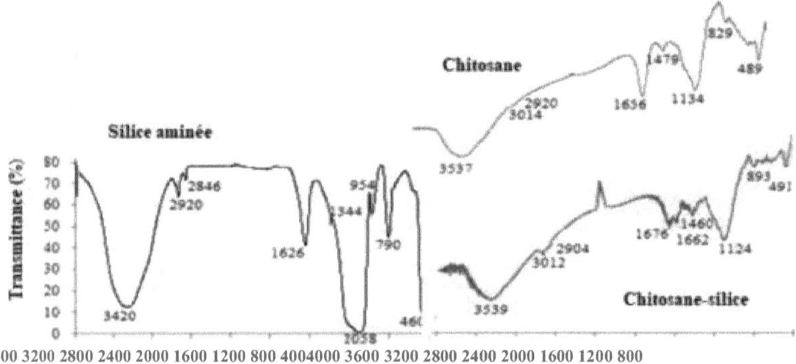

Figure II. 2. IR spectra of amino-silica, pure chitosan and chitosan-silica.

The structure of the chitosan and chitosan-silica beads is porous and rough (Figures II.3,4). These morphological characteristics indicate that these beads could be used for the adsorption of pollutants. I . immobilisation of amiime silica on the surface of chitosan gives more spherical beads (Figure II.4a). This result proves ^rë^^^ of the mechanical resistance of the resulting materials after chemical modification. The silica particles are homogeneously dispersed on the chitosan surface suggesting good adhesion between the biopolymer and the silica.

Figure II. 3. SEM images of unmodified beads: (a) x 1 mm, (b) x 100 and (c) x 200.

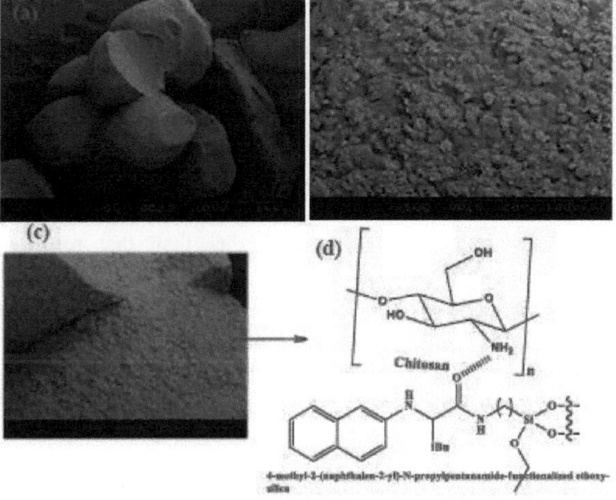

Figure II. 4. SEM images of functionalisedë beads: (a) x 1 mm, (b) x 100 and (c) x 200, and (d) a mechanism showing the interaction between chitosan molecules and aminated silica.

II.3. Application of chitosan beads for dye adsorption

It has been reported that the adsorption mechanism depends on a number of experimental conditions. We study, here, the effects of the paramëtres of pH, contact time, initial dye concentration, and tempërature on 1 adsorption of acid Blue 25 (AB25) and тёЛу^ие blue (BM) dyes.

The adsorbed quantity of the AB25 dye is highest at pH = 5 (Figure II.5a). Under acidic conditions, the amine groups of the biopolymer are protonated, which favours their interaction with the SO_3 groups⁻ of AB25. At higher pH, the adsorption of AB25 on the surface of the beads decreased because in these conditions, there was more OH⁻ in the solution, causing repulsion between the entities [47]. Under alkaline conditions, the amine groups are not protonated and the interaction between AB25 and the adsorbent involves Van Der Waals forces [48]. In the case of BM, the maximum adsorbed quantity was obtained at pH = 6

(Figure II.5b). In fact, the cationic dye dissolved in the water, releasing positive charges. Consequently, under acidic conditions, the surface of the adsorbent carrying positive charges opposes the adsorption of the cationic molecules of BM. As the pH of the solution increases, the surface becomes negatively charged. BM adsorption therefore becomes more important due to the increase in electrostatic forces between the positive charges of the dye and the negative charges of the beads.

Adsorption of the two dyes in the presence of the beads is maximal after 120 min. After 10 min, rapid adsorption of AB25 molecules is achieved (Figure II.5c). This can be explained by the immediate availability of a large number of adsorption sites during this first stage. The difference in adsorption capacity between the anionic dye and the cationic dye (Figure II.5d) is explained by the difference in the binding groups of the two dye molecules and their chemical interactions with the reactive groups of the adsorbents. Indeed, the main characteristic groups of the beads are nitrogen and oxygen atoms. In addition, AB25 is rich in donor atoms in its chemical structure. This makes AB25 molecules better adsorbed than BM molecules.

Figure II.5e, d shows that the adsorption capacity of silica-functionalised beads increases considerably with the initial dye concentration. This behaviour is justified by the number of active sites accessible for adsorption during this phase. At higher dye concentrations, the adsorption capacity slows down and stabilises. This suggests that the interstitial spaces in the adsorbents have become saturated. In fact, adsorption could take place synchronously on the surface and inside the beads. The adsorption of BM also improved using the functionalized chitosan beads, due to the addition of amine groups from the silica amine that are immobilized on the biopolymer. The adsorbed quantity of BM for the functionalised beads (0.65 mg/g wet) is approximately three times higher than that recorded for the non-functionalised beads (0.23 mg/g wet).

The maximum quantity adsorbed is also affected by the change in reaction temperature and decreases as the temperature value increases (FigureII.5g-j). The interaction between the beads, AB25 and BM is therefore exothermic. This indicates that the adsorption mechanism is more favourable at room temperature.

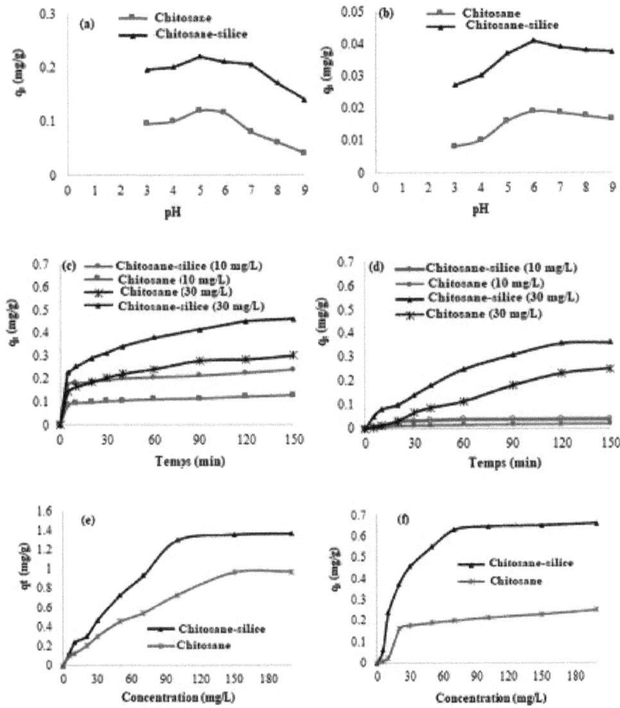

Figure II. 5 (a) Effect of pH on the adsorption of AB25 (C0 = 10 mg/L, T = 20°C), (b) Effect of
pH on the adsorption of BM (C0 = 10 mg/L, T = 20°C), (c) Effect of cinetics on the adsorption of
AB25, (d) Effect of cinetics on the adsorption, (e) Effect of concentration

The adsorption of both dyes was ëtudiëe analysed by the theoretical pseudo-first-order, pseudo-second-order, Elovich and intra-particle diffusion cinetic ëquations [49, 50]. The cinetic paramëtres have been grouped in Tables II.3, 4. The high correlation coefficients (R^2 > 0.99) obtained in the case of AB25 adsorption indicate that the kinetic data are consistent with the pseudo-second-order model. This suggests that the adsorption of AB25 is chemical [49, 50]. On the other hand, the adsorption of BM correlates well with the pseudo first order for non-functionalized chitosan and agrees well with the pseudo second order for functionalized beads. This result can be explained by the difference in the chemical structures of AB25 and BM and also the functional groups present in the chitosan beads. Figure II.1 predicts some kind of probable interaction between chitosan, amine silica, AB25 and BM. Indeed, the chitosan molecules could interact with the silica through the amine groups present in their structures via hydrogen bonds. In the coloured solution, the chitosan-silica complex could interact with the dye of AB25 by hydrogen bonding due to the presence of oxygen and nitrogen in the dye molecule or by electrostatic interaction due to the presence of the anionic sulphonate groups of AB25 and the amine groups of chitosan. The BM molecules could react with the functionalized beads via a hydrogen bond.

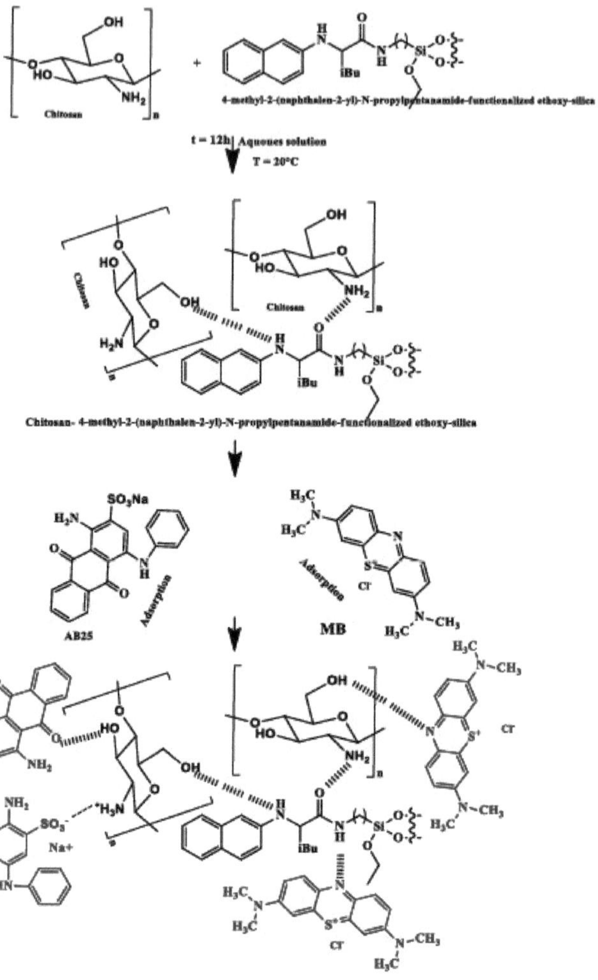

Schema II.1. A probable mechanism representing the interaction between chitosan molecules chitosan molecules, aminated silica, AB25 and MB dyes

To describe the interaction between chitosan beads, AB25 and BM, the data were analysed by Langmuir, Freundlich, Temkin and Dubinin isotherms [51, 52] (tables II.3,4). All values of n are between 1 and 2 indicating that the phënomëne of adsorption of both dyes is favourable [53]. The adsorption energy constants (B), dëterminëed from the Temkin isotherm, decrease with increasing tempërature. This is in line with the decrease in the adsorbed amount of dye with increasing tempërature. This finding is ëalso supported by the negative enthalpy values. The average free energy values, calculated from Dubinin liquidation, range from 74.53 to 223.61 Kj. mol^{-1} . These values are greater than 40 Kj. mol^{-1} suggesting a chemical mechanism [54]. The AS° values are negative, indicating a reduction in the randomness at the solid-solution interface [55].

Table II. 3: Summary of the kinetic constants, Langmuir, Freundlich and Temkin parameters for the adsorption of BM and AB25 (non-functionalised chitosan).

BM

Kinetic equations	Constants	Dye concentration	
		10 mg/L	30 mg/L
Pseudo first order			
	$K1\ (\text{min}^{-1})$	0.008	0.008
	$q(\text{mg-g})^{-1}$	0.017	0.341
	R^2	0.98	0.94
Pseudo-second Order	$K2$	0.788	0.0041
	Q	0.026	0.58
	H	0.0005	0.0014
	R^2	0.88	0.227
Elovich	$o(\text{mg.g}^{-1}.\text{min}^{-1})$	0.0015	0.0067
	$\beta\ (\text{mg.g-1.min-})^{1}$	200	14.28
	R^2	0.99	0.88
Intra-particle diffusion	$K(\text{mg.g}^{1}.\text{min})^{1/2}$	0.001	0.023
	R^2	0.93	0.94

Isotherms	Parameters	Temperature (°)		
		20	40	60
	$q_m\ (\text{mg.g}^{1})$	0.236	0.2	0.185
Langmuir	$K_L\ (\text{L.g}^{-1})$	0.091	0.097	0.06
	R^2	0.99	0.99	0.99
Thermodynamic parameters s	$AH° (\text{KJ mol}^{-1})$	-8.2		
	$AS° (\text{J mol}^{-1})$	-47.17		
	$AG° (\text{KJ mol}^{-1})$	5.62	6.56	7.5
Freundlich	$K_F(\text{L.g})^{-1}$	0.048	0.033	0.028
	n	1.236	1.443	1.32
	R^2	0.74	0.6	0.67
Temkin	B	0.061	0.045	0.044
	$A(\text{L.g})^{-1}$	0.317	0.561	0.394
	R^2	0.86	0.87	0.9
Dubinin	qd	0.185	0.183	0.154
	E	158.11	223.16	223.16
	R^2	0.86	0.99	0.99

AB25

Kinetic equations	Constants	Dye concentration	
		10 mg/L	30 mg/L
Pseudo first order			
	$K_I\ (\text{min})^{-1}$	0.007	0.008
	$q(\text{mg.g})^{-1}$	0.059	0.205
	R^2	0.867	0.963
Pseudo-second Orre	$K2$	1.19	0.242
	Q	0.127	0.317
	H	0.019	0.024
	R^2	0.99	0.99
Elovich	$o(\text{mg.g}^{-1}.\text{min})^{-1}$	2.57	0.136
	$\beta\ (\text{mg.g}^{-1}.\text{min})^{-1}$	90.9	21.27
	R^2	0.96	0.97
Intra-particle diffusion	$K(\text{mg.g}^{1}.\text{min})^{1/2}$	0.007	0.021
	R^2	0.66	0.88

Isotherms	Parameters	Temperature (°)		
		20	40	60
	$q_m(\text{mg.g})^{-1}$	1.7	1.65	1.1
Langmuir	$K_L(\text{L.g})^{-1}$	0.007	0.0043	0.005
	R^2	0.96	0.99	0.98
Thermodynamic parameters	$AH° (\text{KJ mol})^{-1}$	-7.089		
	$AS° (\text{J mol})^{-1}$	-66.25		
	$AG° (\text{KJ mol})^{-1}$	12.32	13.64	14.97
Freundlich	$K_r(\text{L.g})^{-1}$	0.0297	0.0021	0.0026
	n	1.474	0.84	0.912
	R^2	0.99	0.95	0.93
Temkin	B	0.258	0.213	0.157
	$A(\text{L.g})^{-1}$	0.159	0.028	0.127
	R^2	0.9	0.91	0.946
Dubinin	qd	0.33	0.49	0.274
	E	158.11	223.61	158.11
	R^2	0.78	0.54	0.81

Table II. 4: Summary of the kinetic constants, Langmuir, Freundlich and Temkin for the adsorption of BM and AB25 (functionalized chitosan).

BM

Kinetic equations	Constants	Dye concentration	
		10 mg/L	30 mg/L
Pseudo first order			
	$K_I(\text{min})^{-1}$	0.010	0.011
	$q(\text{mg.g})^{-1}$	0.027	0.439
	R^2	0.90	0.96
Pseudo-second order	$K2$	0.767	0.154
	Q	0.05	0.49
	H	0.0019	0.037
	R^2	0.95	0.99
Elovich	$o(\text{mg.g}^{-1}.\text{min})^{-1}$	0.0048	0.0205
	$\beta\ (\text{mg.g}^{-1}.\text{min})^{-1}$	90.9	9.9
	R^2	0.88	0.93
Intra-particle diffusion	$K(\text{mg.g}^{1}.\text{min})^{1/2}$	0.003	0.032
	R^2	0.8	0.98

Isotherms	Parameters	Temperature (°)		
		20	40	60
	$q_m\ (\text{mg.g})^{-1}$	0.725	0.617	0.554
Langmuir	$K_L\ (\text{L. g})^{-1}$	0.06	0.058	0.055
	R^2	0.99	0.99	0.99
Thermodynamic parameters	$AH° (\text{KJ mol})^{-1}$	-1.75		
	$AS° (\text{J mol}^{-}1)$	-29.34		
	$AG° (\text{KJ mol})^{-1}$	6.84	7.42	8.016
Freundlich	$K_F(\text{L.g})^{-1}$	0.045	0.032	0.02
	n	1.766	1.647	1.608
	R^2	0.78	0.76	0.77
Temkin	B	0.166	0.144	0.13
	$A(\text{L.g})^{-1}$	0.417	0.389	0.371
	R^2	0.943	0.934	0.934
Dubinin	qd	0.6	0.48	0.43
	E	223.6	223.6	223.6
	R^2	0.96	0.96	0.956

AB25

Kinetic equations	Constants	Dye concentration	Isotherms	Parameters	Temperature (°)		
		10 mg/L 30 mg/L			20	40	60
Pseudo first order							

Model	Parameter			Model	Parameter			
	K_i (min)$^{-1}$	0.009	0.009	Langmuir	q_m (mg.g^{-1})	2.092	1.43	1.08
	q(mg.g$^{-4)}$	0.109	0.327		K_L(L.g)$^{-1}$	0.012	0.011	0.01
	R^2	0.82	0.97		R^2	0.934	0.989	0.983
Pseudo-second Order	K_2	0.725	0.154	Parameters thermodynamics	$\Delta H°$ (KJ mol)$^{-1}$		-3.68	
	Q	0.237	0.49		$\Delta S°$ (J mol^{-1})		-49.33	
	H	0.04	0.037		$\Delta G°$ (KJ mol)$^{-1}$	10.76	11.75	12.74
	R^2	0.99	0.99		K_r(L.g)$^{-1}$	0.045	0.0324	0.0196
Elovich	o(mg.g^{-1}.min)$^{-1}$	31.65	0.234	Freundlich	n	1.456	1.456	1.372
	$в$ (mg.g^{-1}.min)$^{-1}$	58.82	13.88		R^2	0.974	0.976	0.963
	R^2	0.92	0.97		B	0.387	0.264	0.196
Intra-particle diffusion	K(mg.g^{1}.min)$^{1/2}$	0.013	0.032	Temkin	A(L.g)$^{-1}$	0.172	0.181	0.179
	R^2	0.6	0.88		R^2	0.923	0.958	0.973
				Dubinin	q_d	0.8	0.57	0.43
					E	223.61	223.62	74.53
					R^2	0.63	0.71	0.74

III. Grafting of ethylene diamine and hydrazine onto laurel fibres

III.1 Fibre pre-treatment and grafting

Fibres from collected seeds (Figure II.6) were ёle pretreated according to a procёdё dёcrit dansnos travaux precedents [56]. Indeed, the fibres were impregnated in an aqueous solution (RDB=1:40) containing NaOH (1.5 mL/L, 10M), a non-ionic detergent (4 g/L) and Na2C03 (3 g/L). The mixture was boiled for 90 minutes to remove waxes and impurities. The fibres were then washed in hot and cold water to avoid surface precipitation of impurities, and air dried. Finally, they are washed with distilled water. The purified fibres are treated with a solution of NaOCl (87.5 mL/L, 12°) and Na2C03 (2 g/L) (pH 9-10) for 60 min at room temperature. They were then rinsed with cold water and treated with a solution containing NaHS03 (2 g/L) and H2S04 (0.1 M) at 30°C for 15 min at room temperature. Finally, they were rinsed with distilled water and air-dried.

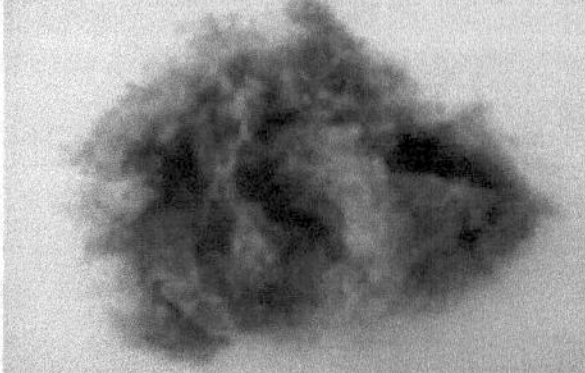

Figure II. 6: Collection and preparation of oleander fibres

Chemical modification of the pretreated fibres was ёle rёalisёe according to a procёdё dёcrit in the literature with some modifications [57]. Briefly, a fibre mass (5.0 g), previously activated at 80°C for 12 h, was kept in a volume of DMF (100 mL). A volume of 17.5 mL of SOCh was then added slowly under continuous stirring at 80°C for 4 h. The cellulose chloride obtained during the reaction was washed with a dilute solution of ammonium hydroxide (0.1 M) and the supernatant after each treatment was removed until the pH reached a neutral value. The suspension was then treated with distilled water to remove the DMF. The fibres were recovered by filtration and dried at room temperature. Amination was carried out at reflux for 4 h using the chlorinated fibres (4.0 g) and a volume of ethylenediamine (20 mL) or hydrazine (20 mL). At the end of the reaction, the ethytene diamine laurel (ED- laurel) and hydrazine laurel (HD- laurel) fibres were separated from the liquid by filtration and washed with water to remove excës of reagent, and finally dried in vacuo.

III.2 Characterisation of grafted fibres

Elemental analysis is used to determine the number of graft groups. Here, we measured the percentages of carbon (C%), hydrogen (H%) and nitrogen (N%) for the fibres studied (Table II.5).The number of coordination sites per gram for each sample, Ca (mmol. g^{-1}), was calculated on the basis of the value of the nitrogen content (Equation 2)[58]. The corresponding values for diamine laurel ethytene and hydrazine laurel are, respectively, equal to 0.847 and 0.588 mmol. g^{-1} :

$$Ca\ (mmol/g) = 3.33 \times \frac{(\%N)}{M(N)} (2)$$

Or :

% N is the nitrogen content

M (N) is the molecular weight of nitrogen.

Table II. 5: Summary of carbon, hydrogen and nitrogen percentages for the fibres fibres studied.

Fibres studied	C (%)	H (%)	N (%)	Ca (mmol.g)$^{-1}$
Bleached fibres	44.216	6.229	-	-
ED-laurier	46.006	6.135	3.558	0.847
HD-laurier	40.130	5.710	2.473	0.588

The IR spectra of the fibres studied are shown in Figure II.7. The crude fibres (Figure II.7b) show a broad band at around 3350 cm-1 which corresponds to the OH group [59]. The band at 889 cm^{-1} (amorphous region of cellulose) is attributed to 0-glucosidic bonds [60]. The band at 1024 cn^{-1} is attributed to the C-OH stretching vibration (secondary alcohol v C-O) [61]. Symmetric CH bending of methoxyl groups was observed at 1367 cm^{-1} [62]. Symmetrical CH2 bending was observed at 1448 cm^{-1} (cellulose crystal structure). The two bands at about 2881 and 2835 cm^{-1} are related to asymmetric and symmetric methyl and methylene stretching groups [63]. The bands at 1591-1589 and 1278 cm^{-1} are attributed to C = C, C-O stretching or bending vibrations of the lignin groups [64]. The band at 1765 cm^{-1} , in the spectrum of raw fibres, is due to the C = O bond, which is a characteristic group of lignin and hemicelluloses. From the IR analyses of raw (Figure II.7b) and bleached (Figure II.7a) laurel fibres, it was observed that the bands at around 1765 (C = O bond) and 1200-1300 (aromatic vibration), characteristic of hemicelluloses and lignin, disappeared after pre-treatment. These findings were observed in the study by Abraham et al, [65] when studying the extraction of cellulose from lignocellulosic fibres.

By comparing the IR spectrum of laurel fibres with that of ED-laurier (Figure II.7c) and HD-laurier (Figure II.7d), some differences were recorded. In fact, the band at 3350 cm^{-1} (OH stretching groups) shifted to 3362 cm^{-1} revealing the addition of NH groups from ethylene diamine and hydrazine reagents. Two new peaks also appeared at 1618 and 1541 cm^{-1} which are attributed to the deformation vibration of NH in the primary amine (-NH2) and N-H in the secondary amine (-NH) [66].

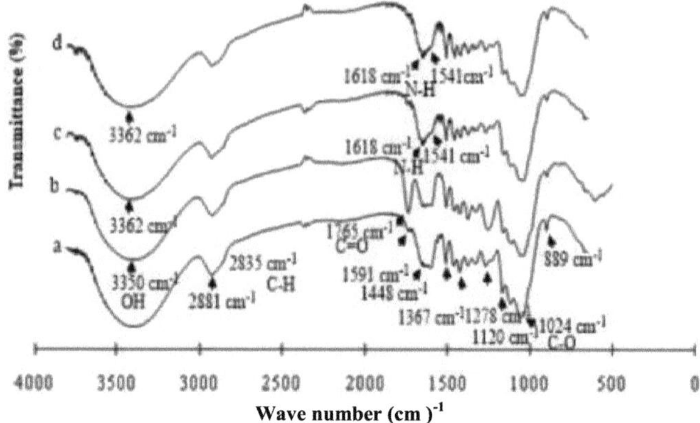

Figure II. 7: IR spectra of: (a) bleached fibres, (b) raw fibres, (c) ED-laurier, and (d) ED-laurier.

HD-laurier

The morphological characteristics of the fibres have ële ëtudiëes (Figure II.8). Figure II.8a shows images of ungrafted fibres where the surfaces are smooth, minimally twisted and clean with numerous small striations. Figures II.8b, c show images of grafted fibres. The aminated fibres have become more deformed than the unmodified fibres. This deformation is greater in the case of fibres modified with ethylenediamine. This is explained by the high reactivity of the amine groups present in the structure of this reagent towards the hydroxyl groups of the cellulose.

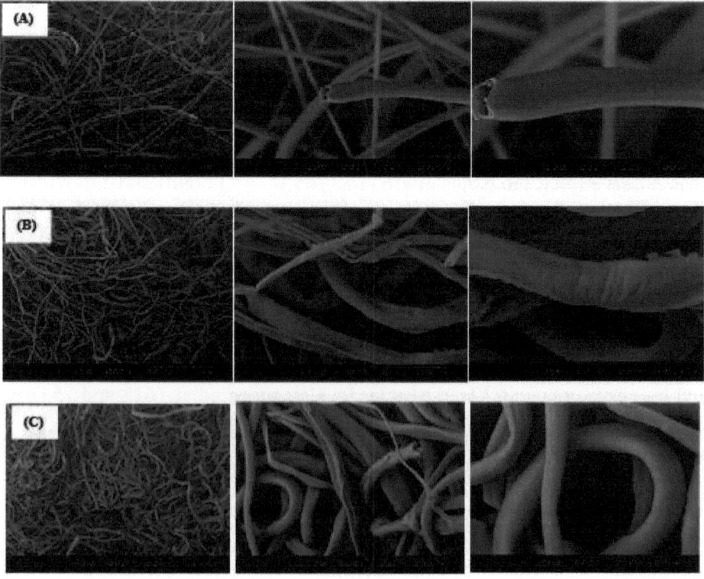

Figure II. 8. SEM images of: (A) laurel fibres, (B) ED-laurier, and (C) HD-laurier observed at various magnifications (x 30, x 200 and x 500)

III.3. Application of grafted fibres for the adsorption of anionic dyes

Cellulosic fibres have poor affinity to anionic dyes due to electrostatic repulsion between the negative hydroxyl groups (OH^-) present on the fibre surface and the sulphonate groups (SO_3^-) of the dye molecules. To improve the affinity of laurel fibres, we proposed grafting with ethylenediamine and hydrazine. In this section, we studied the influence of experimental conditions on the adsorption of two acid dyes: Acid Blue 161 (AB161) and Acid Red 183 (AR183) (Table II.6).

Table II. 6: Chemical structures and physical characteristics of the dyes studied

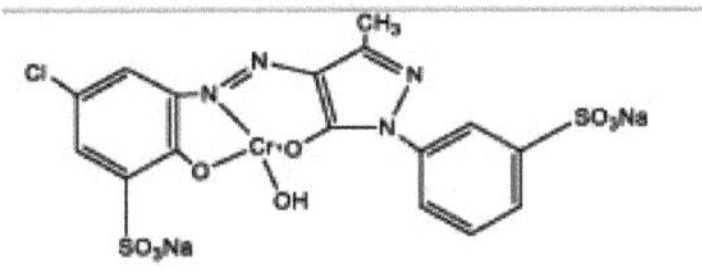

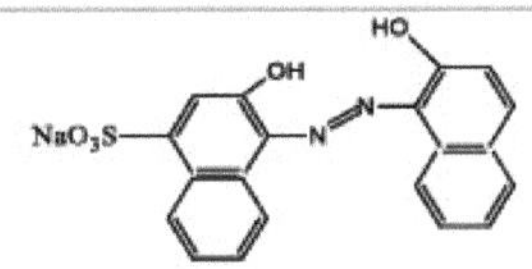

Red 183 acid (AR183) ^max in water = 495 nm Molar mass (g/mol) = 599.83
Acid Blue 161(AB161)
^max in water = 609 nm Molar mass (g/mol) = 394.4

The amount of AB161 adsorbed is significantly affected by varying pH values (Figure II.9a). For example, when the pH varies from 4 to 8, qt decreases from 5.5 $mg.g^{-1}$ to 1.5 $mg.g^{-1}$. Increasing the pH value from 3 to 4 increases the adsorption capacity of AB161, while increasing the pH value from 4 to 8 leads to a decrease in the adsorbed quantity. The relatively high adsorbed quantity at pH values below 4 can be explained by electrostatic interactions between the positively charged surface of the adsorbent and the negatively charged AB161 molecules. The large amount of dye adsorbed at pH 4 for ED-laurier fibres could be attributed to the ëlectrostatic interaction between the proton amine groups of ethylene diamine and the anionic groups of the dye molecules formed by dissociation under acidic conditions. The evolution of the adsorbed amounts of AR183 and AB161 is shown in FigureII.9b, c. During the first 30 minutes, adsorption of both dyes is rapid, as there were more sites on the surface of the adsorbent available at the start of the process. It then proceeds at a slower rate, as there are few active sites on the surface of the adsorbents, and reaches saturation at 90 min.

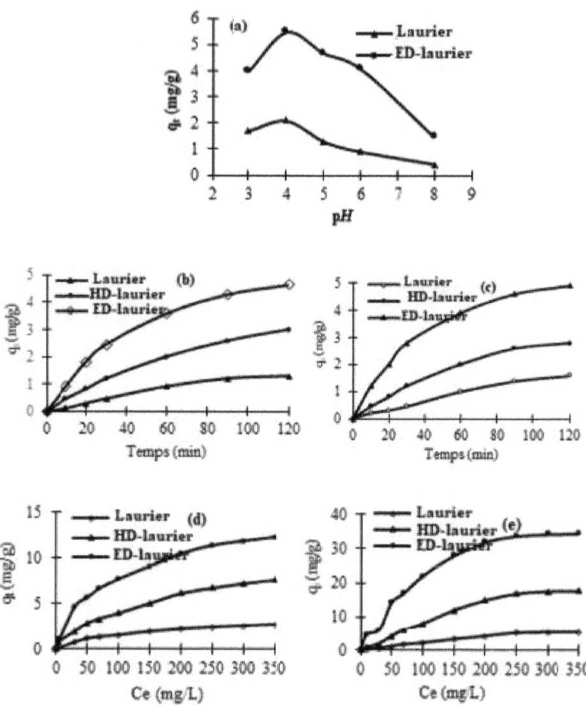

Figure II. 9: (a) Effect of pH (C0 = 30 mg / L, T = 25°C, t = 60 min), (b) Effect of time : (C0 = 30 mg / L, T = 25°C, pH = 6), Effect of dye concentration: (d) AR183 and (e) AB 161 (T = 25°C, t = 3h, pH = 6)

FigureII.9d, e shows Involution of the adsorbed quantile of ëtudiës adsorbates as a function of dye concentration. The highest adsorbed quantity (35 mg g^{-1}) is obtained with the dye AB161 using ED-laurier. It is approximately 12.2 mg g^{-1} for dye AR183. The ED-laurier adsorbent proved more effective than HD-laurier. In fact, the quantity adsorbed for this second adsorbent is reduced to 50% compared with that for ED-laurier. Moreover, these quantities do not exceed, respectively, 2.7 and 5.2 mg g^{-1} for AR183 and AB161 using raw fibres as adsorbents, under the same conditions. The adsorbed amount of AB161 is relatively higher using ED-laurier as adsorbent compared to AR183. This difference in adsorbed quantity is explained by the high molecular weight of AR183 molecules, which prevents them from being absorbed by diffusion mechanisms. The adsorbed quantities of AR183 increase from 12.2 to 16.25 mg.g^{-1} when the temperature increases from 25°C to 60°C, in the case of ED-laurier (Figure II.10). This could be explained by the following reasons: firstly, the pore size of the adsorbent particles increasesat high temperatures [67]. Secondly, the number of adsorption sites increases due to the breaking of certain internal bonds [68]. In the case of the use of unmodified laurel fibres as adsorbent, the adsorbed amount of AR183 decreases with increasing latemperature.

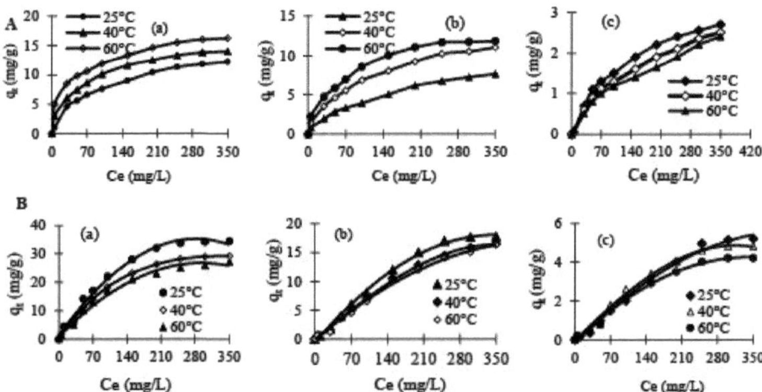

Figure II. 10 Effect of temperature: (A) AR183 and (B) AB161 using: (a) ED-laurel, (b) HD-laurel, and (c) Laurel fibres (t = 3h, pH = 6)

The calculated cinetic parameters have ële rësumës in Table II.7. The results indicate that the pseudo-first order fits the experimental data better with regression coefficients $R_2 > 0.96$. Furthermore, the qe values calculated for this kinetic model show good agreement with the experimental ones (low SSE values). The process could therefore be considered physical. The isotherm parameters are shown in Table II.8. The relatively low correlation coefficients show that the Langmuir isotherm does not agree with the experimental data. On the other hand, the Freundlich equation correlates fairly well with the experimental data, with fairly high correlation coefficients ($R^2 > 0.91$). This suggests that the adsorption phenomenon can occur in heterogeneous adsorption sites. The fibres studied are moderate and sometimes good adsorbents for AR183 (n> 2) [69]. The E values calculated from the Redushkevich equation are in most cases less than 40 kJ mol^{-1} confirming that the adsorption of AB161 and AR183 is physical [69].

Table II. 7. Cinetic parameters for the adsorption of AR183 and AB161.

Fibres	First-order nickname				Pseudo second-order				Elovich			Intra-particle diffusion	
	K1	qe	R^2	SSE	K2	qe	R^2	SSE	aB		R^2	K1	R^2
AR183													
Laurel	0.006	1.711	0.99	0.079	0.002	2.889	0.22	0.267	0.051	1.952	0.973	0.133	0.943
HD-laurel	0.008	3.643	0.99	0.077	0.008	4.206	0.69	0.199	0.129	0.942	0.97	0.293	0.973
ED-laurel	0.01	5.047	0.99	0.057	0.005	5.78	0.87	0.187	0.262	0.644	0.998	0.457	0.984
AB161													
Laurel	0.01	2.083	0.96	0.08	0.005	2.471	0.46	0.145	0.063	3.075	0.739	0.156	0.93
HD-laurel	0.012	3.533	0.98	0.12	2.227	3.905	0.73	0.184	0.264	1.655	0.837	0.281	0.973
ED-laurel	0.014	5.599	0.98	0.116	0.006	5.81	0.91	0.15	0.611	0.936	0.92	0.476	0.985

Table II. 8 Langmuir, Freundlich, Temkin and Dubinin parameters for the adsorption of AR183 and AB161

Langmuir	Freundlich	Temkin	Dubinin

T°C	q_m	KlR^2	SSE	$KfnR^2$	BAt	R^2	q_m	ER SSE^2
				AR183				
				Laurel fibres				
25	5.010	.004 0.6	0.231 0.000151	.	0.	650.136 0.95 1.6214		.43 0.890
			120.91			.108		
40	4.590	.003 0.64	0.209 0.000145 1.165 0.92		0.590	.125 0.931	.397 23.57 0.880	.11
60	3.	080.006 0.96	0.068 0.00071	.429 0.98	0.	530.125 0.89 1.233 40.82 0.770		.117
				HD-laurier				
25	9.	460.009 0.94	0.186 0.0701	.813 0.99	1.710	.153 0.89 4.288 20.41 0.		640.79
40	12.97 0.013 0.97		0.197 0.3802	.024 0.99	2.	420.20 0.94 6.932 35.35 0.690		.7
60	13.87 0.018 0.98		0.207 1.4332	.417 0.98	2.	540.299 0.95 8.468	500.	680.78
				ED-laurier				
25	14.79 0.012 0.99		0.259 0. 21.720 0.96		2.745 0.202	0.97 8. 118.26 0.842 0.41		
40	15.75 0.021 0.99		0.175 3.452	.687 0.99	2.806 0.389	0.97 10.51 40.82 0.		70.35
60	17.64 0.027 0.99		0.1414 .763	.542 0.99	2.799 0.836	0.97 12.73 70.71 0.		680.351
				AB161				
				Laurel fibres				
25	34.36 0.0006 0.59		2.194 0.010 1.150 0.98 1.329 0.096 0.845 2.34315			.811 0.629 0.28		
40	11.16 0.0020	.57	0.637 0.009 1.184 0.98 1.143 0.101 0.832	.12526		.726 0.658 0.27		
60	100.0020	.65	0.580 .0111	.220 0.95 0.988 0.101 0.825 1.77740		.824 0.537 0.24		
				HD-laurier				
25	56.820 .0020	.49	3. 90.136 1.149 0.98 4. 750.089 0.842 8.29216		.660	.578 2.62		
40	55.860 .0010	.3	3.940 .1311	.183 0.98 4. 210.087 0.838 7.20326		.726 0.567 2.20		
60	52. 360.0010	.22	3. 60.1411	.219 0.95 4.050 .086 0.824 6.70140		.824 0.493 2.052		
				ED-laurier				
25	50. 500.0080	.85	1. 62.337 1.756 0.93 8.486 0.157 0.8920 .999 21. 320.491 .35					
40	42.918 0.0070	.98	1. 370.4711	.493 0.98 7.359 0.140 0.9318		.159 28.867 0.744 1.1		
60	39.525 0.0070	.96	1. 230.249 1.424 0.97 6.760 0.132 0.926 15.921 40.824 0.711			.13		

The thermodynamic parameters were ë1.ë calculated (Table II.9). The negative value of the enthalpy confirms that the adsorption of AB161 on the grafted fibres is an exothermic process. The adsorption of AR183 is endothermic for the grafted fibres, which could be explained by the release of the Cr element (i.e. the breaking of Cr-L or Cr-X bonds versus the formation of electrostatic bonds).The negative value of AS* indicates a decrease in the disorder of the system at the grafted fibre-AB161 interface. For AR183, positive values of AS* indicate increased dësorder at the solid/solution interface. Positive AG* values mean that adsorption of the two dyes studied is non-spontaneous.

Table II. 9: Thermodynamic parameters for the adsorption of AR183 and AB161.

Fibres		AH* (kJ.mol)$^{-1}$	AS*(J.mol)$^{-1}$	AG* (kJ.mol)$^{-1}$		
	T (°C)	25 4060	254060	25	40	60
Laurel	AR183	12.737	-4.298	13.748	14.621	13.916
	AB161	21.275	13.21	18.265	15.357	17.744
HD-laurier	AR183	16.353	15.755	11.688	11.380	11.135
	AB161	-3.999	-66.985	15.886	17.113	18.235
ED-laurier	AR183	18.573	26.305	11.429	10.061	9.951
	AB161	-3.293	-51.455	12.059	12.773	13.859

IV. Amination of almond shells
IV.1 Preparation of adsorbents from almond shells

The almond shells (Figure II.11) were ële harvested in the Sidi Bouzid region of Tunisia in summer (July-August). First of all, they were rinsed several times with distilled water to reduce the impurities deposited on the surface, sëchëed and then ground with a mortar followed by electric grinding to obtain finer powders.

Figure II. 11: (a) Almond tree, (b) fresh almond fruit and (c) almond fruit after ripening.

The almond shell powders were treated by two cationisation processes using two different amine reagents. The first used chitosan (0.05, 0.1, 0.3 and 0.5%). The second used a dimethy-diallyl-ammonium-chloride-diallylamine copolymer (0.1, 0.3, 0.5, 2, 3 and 5%) (Figure II.12).

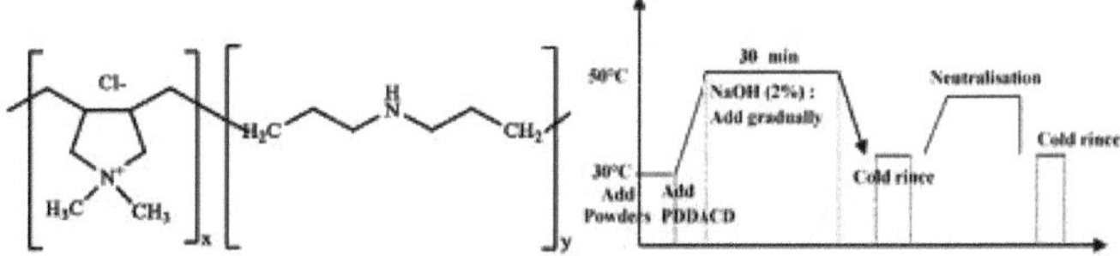

Figure II. 12: (a) Chemical structure of dimethy-diallyl-ammonium-chloride-diallylamine co-polymer and (b) cationisation processes.

IV.2 Characterisation of almond shells

SEM images of the unmodified and modified shells have ële given in Figure II.13. The surface is rough and porous resembling a honeycomb. The structure is macroporous where pore diameters vary from 26.51 to 33.44 |im and microporous having pores varying from 0.78 to 1.35 ^m. For the almond shell powders modified with the copolymer (Figure II.13b), no change was observed compared to the raw materials (Figure II.13a). Furthermore, the application of chitosan to the surface of the shells leads to the clogging of the pores (Figure II.13c). This imparts a certain amount of stiffness to the materials.

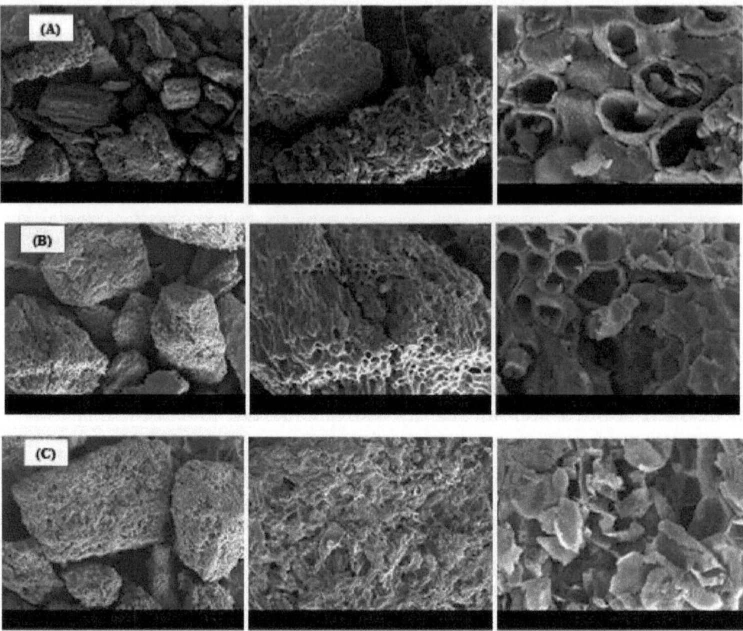

Figure II. 13. SEM images of: (a) raw almond shells, (b) almond-copolymber, and (c) almond-chitosan (x50, x 125, and x500).

The IR spectra of the ëtudiës are shown in Figure II.14. The spectra of the crude almond shells show the presence of a band at about 3316 cm^{-1} which corresponds to the OH group [60]. The peak at 867 cm^{-1} is attributed to 0-glucosidic bonds [60]. The band at 1040 cm^{-1} is attributed to the C-OH stretching vibration [61]. The bands at around 2865 and 2957 cm^{-1} are associated with OCH3 groups [61]. The peaks at 1700 (C=O bond) and 1224 cm^{-1} are groups characteristic of lignin and hemicelluloses [64]. The bands at 1580 and 1286 cm^{-1} are attributed to the C=C and C-O groups of lignin. The spectrum of functionalised almond shells shows a shift from the band at 3316 cm^{-1} to 3363 cm^{-1} . This confirms the addition of chitosan and copolymer amine groups to the almond shell surface. For the almond-copolymer spectrum, a new peak appears at 1603 cm^{-1} which is attributed to the N-H group [64].

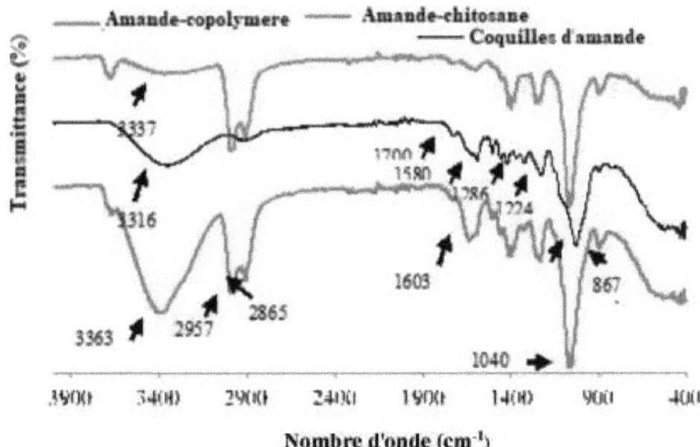

Figure II. 14. IR spectra of almond, almond-chitosan and almond-copolymershells.

IV.3. Application of almond shells for the adsorption of anionic and cationic dyes

Contact between the adsorbate and the adsorbent leads to a physical and/or chemical interaction. This depends on the structure of the materials. Indeed, the resulting adsorbed quantity dëpend on several physicochemical paramëtres. In this section, we ёШёюпз! 'effect of pH, contact time, temperature, and initial dye concentration.

The adsorbed quantity of AB25 (Figure II.16a) is maximal at pH = 2 and 3, respectively, for the unmodified almond, almond-chitosan, and almond-copolymerëre shells. Above these values, the qt values decrease. This result can be explained by the fact that under acidic conditions, the H3O ions$^+$ and the proton amine groups of the chitosan or copolymëre interacted with the sulphonic groups of the AB25 molecules. On the other hand, under alkaline conditions, the presence of OH⁻ ions caused repulsion between the dye molecules and the adsorbent. In the following adsorption experiments, the value of pH = 4 was ëtë chosen as the optimum value in order to avoid the degradation of the cellulose materials studied under strongly acidic conditions. For the adsorption of BM on the surface of almond shells (Figure II.16b), a significant increase in the adsorption capacity of the adsorbent was observed when the pH value increased from 2 to 6. Whereas above the value 6, the adsorbed quantity is legërementally affected by the change in pH value. The rather low adsorbed amounts of BM at acidic pH could be due to the presence of an excës of H$^+$ ions competing with the dye cations.

Scheme II.3 predicts a likely mëcanism of interaction between the cellulosicmatërials studied and the dyes from AB25 and BM.

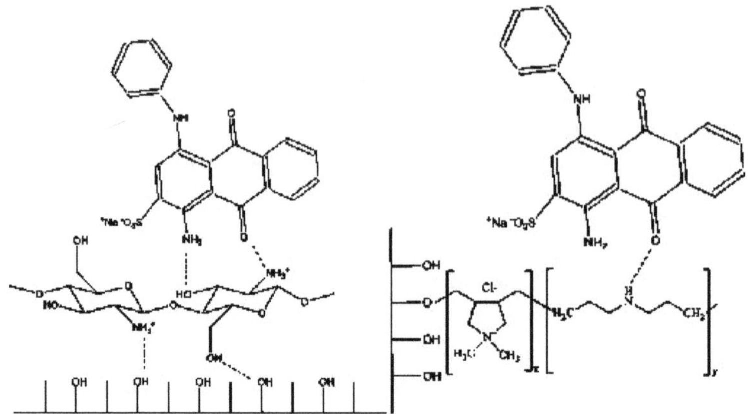

Figure II.3: Probable mechanism of interaction between: (a) almond shells, chitosan and AB25 and (b) almond
shells, PDDCD and AB25.

The variation in the adsorbed quantity of AB25 and BM as a function of contact time for the materials studied using different doses of chitosan and copolymer is shown in Figure II.15. For all the compounds, it can be seen that the quantity adsorbed is greater at the start of the process. AB25 equilibrium (Figure II.15c, d) is obtained at about 60 and 30 minutes, respectively, for the almond-chitosan and almond-copolymer materials. This equilibrium is reached at about 60 min for the BM dye. The amount of AB25 adsorbed increases as the dose of each cationic reagent increases. This is due to the presence of amine groups on the surface of almond-chitosan and almond-copolymer. The optimum doses of cationic reagents for AB25 adsorption are equal to 0.5% and 5%, respectively, for chitosan and copolymer. For BM (Figure II.15e), the amount of dye adsorbed on the surface of the almond shells is explained by the presence of reactive hydroxyl groups.

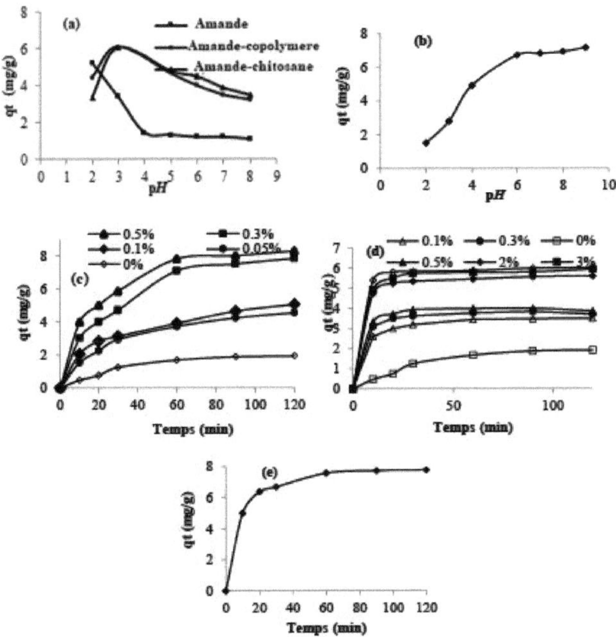

Figure II. 15: Effect of pH value on the adsorption of: (a) AB25 and (b) BM on the surface of almond

shells, almond-chitosan and almond-copolymere (t = 30 min, T = 25°C, C0 =

30 mg/L), Effect of time on the adsorption of AB25 using: (c) almond-chitosan.

Figure II.16 ëlucidates the effect of tempërature on the adsorption of AB25 and BM in the presence of almond shell, almond-chitosan and almond-copolymëre. The interaction between the ëtudiës matërials and AB25 (Figure II.16a-c) is exothermic or the tempërature decreases with increasing tempërature value. The decrease in the quantity adsorbed as a function of temperature then indicates that the interaction established between the adsorbents and AB25 are reversible. The adsorbed quantity at equilibrium, at 60°C, for almond-chitosan is equal to 20 mg.g^{-1} . This capacity is slightly higher for almond-copolymer which is equal to 25 mg.g^{-1} . Furthermore, under the same conditions, the adsorbed quantity of AB25 does not exceed 2.4 mg.g^{-1} for unmodified almond shells. This difference in adsorption capacity proves the contribution of cationic sites to the cellulose surface. The adsorption phenomenon is endothermic in the case of BM, where a slight increase in the quantity adsorbed was observed. At 60°C, the adsorbed quantity is 86 mg.g^{-1} and it is approximately 84.9 mg.g^{-1} at 25°C.

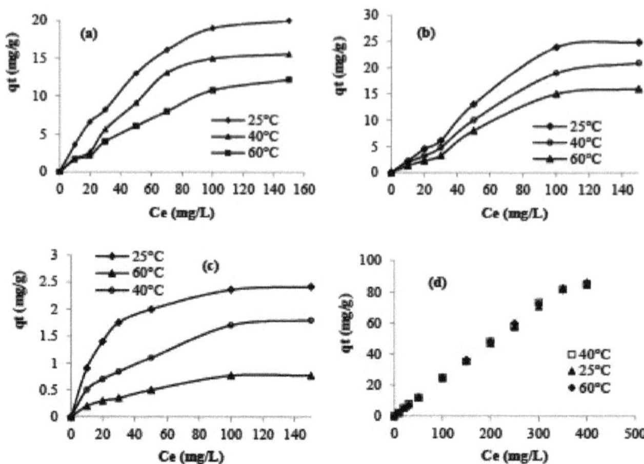

Figure II. 16. Effect of temperature on adsorption of AB25: (a) almond-chitosan, (b) almond-copolymere, (c) almond shells, and (d) MB using almond shells.

The evaluation of the kinetic models is controlled by the values of the regression coefficients R^2 and SSE (Table II.10). For the adsorption of AB25 using almond-chitosan as adsorbent, the values of R^2 are found to be larger, along the pseudo-second order equation. However, the lowest SSE values were recorded for the pseudo-first order. In general, there is a combination of adsorption on the surface and inside the pores. Indeed, the pseudo-first-order equation may describe only the initial phase of the adsorption phenomenon and the adsorption data may deviate from the curve. It has been suggested that the first-order model does not fit well over the entire contact time interval and is generally applicable during the initial stage of adsorption processes. The non-linearity suggests the inability of the pseudo-first-order model to interpret the data. The pseudo-second order equation indicates that the mechanism is chemical.

The intra-particle diffusion model is used to identify the mechanism involved in the adsorption process. It assumes that intra-particle diffusion is velocity-controlled, which is generally the case for well-mixed solutions. The applicability of this model requires that the trace of qt with respect to $t^\wedge$ is linear; if it passes through the origin, intra-particle diffusion is the only rate-controlling step. When the lines do not pass through the origin, this indicates that intra-particle diffusion is not the only rate-controlling step, but that more than one kinetic process is involved in the adsorption process. Here, the straight lines of the kinetic data for the intra-particle diffusion equation pass the origin. This suggests that the diffusion step is significant in this case. Referring to the R values[2] of this equation, we observe that this model is less suitable than the pseudo-first order. Thus, adsorption is predominant at the surface of the adsorbent but remains secondary inside the pores. In fact, the rapid adsorption at the beginning is linked to surface adsorption, then adsorption takes place slowly inside the pores.

In addition, the values of R^2 are high for the Elovich equation, suggesting chemical adsorption with heterogeneous pores on the adsorbent surface. The adsorption process using almond-copolymer solves the pseudo-second order equation. This was confirmed by the low SSE values (Table II.10). For this adsorbent, the intra-particle diffusion lines do not pass the origin. Consequently, this step is not involved in the adsorption process.

BM adsorption on the surface of unmodified almond shells follows second-order liquation $(0.99 < R^2 ; 0.073 < SSE)$. This confirms that BM adsorption is chemical.

Table II. 10: Kinetic parameters for adsorption of AB25 and BM using almond, almond-chitosan and almond-copolymer shells.

AB25 (30 mg/L)	Pseudo-first-order				Pseudo-second order				Elovich			Intra-particle diffusion	
	Ki	qe	R^2	SSE	K2	qe	R	SSE²	a	e	R^2	Ki	R^2
Almond shells													
AB25	0.013	1.9	0.99	0.01	0.00	2		0.13	0.73	0.69	0.97	0.192	0.95
	0.014	9	4	3	8	.7	0.97	1	1.78	0.35	4	0.453	9
	6	4.8	0.96	0.05	0.00	6		0.17			0.99	3	0.96
		7	3	5	6						7		9
Almond chitosan													
0.05						5.56	0 .998						
0.1%													
	0.014	5.2	0.93	0.04	0.00	5.86	0 .990	0.13	2.07	0.36	0.98	0.501	0.94
	4	9	0	1	7			8		6	9		9
0.3%													
	0.007	6.6	0.97	0.18	0.00	9		0.28	3.28	0.2	0.97	0.816	0.88
	2	9	9	6	4	.52	0.9	5		2	4		7
						92							
0.5%													
	0.01587	.08	0.96	0.19	0.00	9.		0.18	3.94	0.23	0.97	0.885	0.88
Almond-copolymer		4	9	6		390.99				4	3	1	6
		7											
0.1%													
	0.013	1.3	0.85	0.35	0.07	3.		0.02	7.92	1.2	0.93	0.27	0.69
	3	7	2		1	620.99		0		1			4
						9							
0.3%													
	0.010	0.9	0.57	0.45	0.17	3.81	0 .998	0.01	63.17	1.75	0.79	0.283	0.60
	1	9	2	3				6			6		4
0.5%													
	0.009	0.7	0.43	0.52	0.36	3.		0.01	196.7	1.99	0.65	0.291	0.57
	8	4	8	1		950.99		0		1			5
						8							
2%													
	0.013	1.5	0.79	0.68	0.09	5.		0.01	262.0	1.41	0.88	0.411	0.59

	KL	qmL	R²	SSE	KF	nF	R²	BT	AT	R²	q	E	R²	SSE	AH	AS	AG°
	3	0	7	3				680.999	3						4		0
0.3%	0.0128	1.59	0.753	0.725	0.08	5.990	.999	0.008	138.6	1.18	0.818	0.437	0.595				
5%	0.0129	1.15	0.7	0.808	0.13	6.050	.9990	.008	1567	22.06	0.830	0.43	0.546				
Almond shells							9										
BM	0.0174	4.32	0.960	0.573	0.0198		200.999		0.0737		640.	390.938		0.640	0.770		0

The parameters calculated from the Langmuir, Freundlich, Temkin and Dubinin-Redushkevich isotherms are summarised in Table II.11.

Table II. 11: Langmuir, Freundlich, Temkin, Dubinin-Redushkevich constants and thermodynamic parameters for the adsorption of AB25 and BM.

	Langmuir				Freundlich			Temkin			Dubinin-Redushkevich				Parameters		
T	KL	qmL	R²	SSE	KF	nF	R²	BT	AT	R²	q	E	R²	SSE	AH kJ.mol⁻¹	AS J.mol⁻¹	AG° kJ.mol⁻¹
AB25 : Almond-0.3%chitosan																	
25	0.22	29.9	0.75	2.53	1.14	0.68	0.97	15.33	0.43	0.97	3.21	91.28	0.781	2.38			13.77
40	0.10	48.3	0.42	4.67	0.52	1.1	0.94	13.73	0.37	0.94	2.84	40.82	0.846	1.82	-12	46.28	14.47
60	0.07	31.2	0.97	1.6	0.53	1.23	0.97	9.72	0.373	0.94	2.38	50	0.798	1.4			15.39
AB25 : Almond-5%copolymer																	
25	0.06	135.13	0.429	19.85	0.56	1.05	0.97	21.89	0.35	0.93	3.21	74.5	0.7	3.62			10.07
40	0.04	133.33	0.23	18.72	0.5	1.05	0.98	18.05	0.34	0.92	2.87	79.05	0.64	3.02	-7.04	-33.83	10.58
60	0.03	123.45	0.343	16.41	0.39	0.98	0.97	14.23	0.39	0.92	2.55	84.5	0.64	2.24			11.25
AB25:Almond																	
25	45.52	1.05	0.974	0.04	0.71	2.82	0.91	1.31	0.79	0.97	1.39	111.8	0.895	0.17			25.5
40	8.48	2.39	0.977	0.09	0.44	2.01	0.99	1.19	0.5	0.958	1.11	111.8	0.66	0.11	-29.5	-85.7	26.79
60	2.49	2.74	0.999	0.05	0.29	1.89	0.981	0.54	0.48	0.955	1.27	111.8	0.69	0.08			28.5
BM: almond																	
25	0.006	625	0.211	49	0.54	1.01	0.997	53.24	0.282	0.868	4.75	70.7	0.621	7.28			13.74
40	0.008	454.5	0.473	33.59	0.53	1	0.996	53.68	0.282	0.870	4.76	70.7	0.625	7.29	4.08	-4.6	14.43
60	0.009	384.6	0.517	27.2	0.52	0.98	0.997	54.28	0.281	0.874	4.74	70.7	0.6	7.38			15.35

The correlation coefficients indicate that the Langmuir and Freundlich ëquations are satisfactory. In detail, by looking at the regression coefficients, we observed that Freundlich liquation is more appropriateëe to dëcribe the adsorption of AB25 using almond-chitosan and almond-copolymëre as adsorbents. On the contrary, the adsorption of AB25 using almond shellsuit Langmuir liquation. This can be interpreted by the homogeneous and heterogeneous distribution of active sites on the almond shell surface. Based on the values of n, the materials studied are moderate adsorbents and sometimes good for the adsorption of AB25 (n> 2). The values of E, calculated from the Redushkevich equation, are greater than 40 KJ mol⁻¹ confirming that the adsorption of AB25 and MB is chemical. For AB25, the negative enthalpy values confirm that the interaction between AB25, almond shells, almond-chitosan and almond-copolymer is exothermic. For BM, the process is endothermic. Positive values of AG° indicate a non-spontaneous reactionev Negative values of AS° suggest a decrease in dësorder during adsorption.

V . Cationisation of cellulose membranes obtained from dates

V .1. Preparation of cellulose membranes

The cellulose membranes were ële obtained from dates (Figure II.17). They were first washed with water to remove surface impurities and then dried at room tempërature. They underwent an additional cationising treatment using dimëthy-diallyl-ammonium-chloride-diallylamine copolymer.

Figure II. 17: Date collection: (a) fruit, (b) dates, and (c) residual membranes.

V . 2. Characterisation of cellulose membranes

The IR spectra of the raw and functionalised membranes are shown in Figure II.18. The crude membranes show a band at about 3265 cm^{-1} which corresponds to the OH group. The peak at 869 cm^{-1} is attributed to B-glucosidic bonds. The band at 1020 cm^{-1} is attributed to C-OH stretching vibration. The symmëtrical CH bending of the mëthoxyl groups has ëlë observed at 1365 cm^{-1} . The two bands at 2885 and 2958 cm^{-1} are ^ to O-CH3 groups. The peaks at 1720 (C = O bond) and 1224 cm^{-1} are characteristic of lignin and hëmicelluloses. The bands at 1589 and 1286 cm^{-1} are attributable to the C = C and C-O stretching vibrations of the lignin groups.The spectrum of the functionalised membrane shows a shift from the band at 3265 cm^{-1} to 3312 cm^{-1} . This confirms the addition of amine groups to thecopolymëre. The peaks at 1427 and 1574 cm^{-1} are attributed, respectively, to the C-N and N- H stretching vibration of the secondary amine. These results demonstrate the interaction between the copolymer and the cellulose.

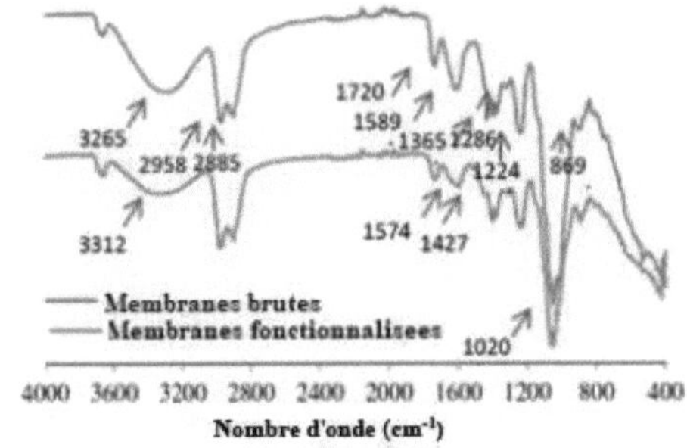

Figure II. 18. IR spectra of raw and functionalised membranes.

The morphological characteristics of the membranes are shown in Figure II.19. Figure II.19a shows that the surface of the unmodified membranes exhibits parallel, shallow grooves. For images of the functionalized membranes (Figure II.19b), there is no clear change in surface morphology. The cationised surfaces are legërementally smoother than the unmodified ones. Cationisation does not alter the physical structure of the fibre. This is an advantage over the treatment of materials with polymers (as in the case of chitosan), which produces a degree of rigidity.

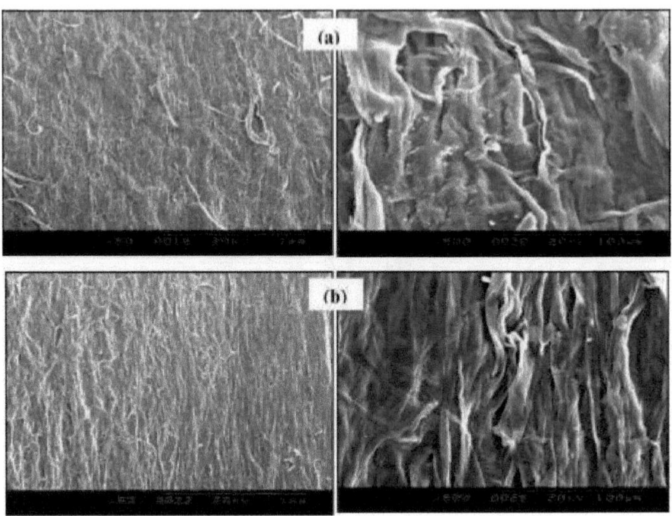

Figure II. 19. SEM images of membranes: (a) raw and (b) functionalised (x50 and x500).

V.3. Application of cellulose membranes for the adsorption of acid, reactive, direct and cationic dyes

The interaction between the cellulose chains of the membranes and the ëtudiës dyes (Table II.12) is conditioned not only by the presence of functional groups on the surface of the adsorbent or adsorbate but also by several experimental parameters such as pH, contact time, initial concentration of the adsorbate, temperature, etc. In fact, the membrane chains could interact with the BM molecules by hydrogen bonding thanks to the presence of nitrogen atoms and hydroxyl groups. In fact, the membrane chains could interact with the BM molecules by hydrogëne bonding thanks to the presence of nitrogen atoms and hydroxyl groups. After cationisation, the copolymer added to the surface of the cellulose chains could react with the NBB anionic dye, for example, via ionic interactions between the N^+ and so_3^- groups.

Table II. 12: Chemical structures and physical characteristics of the dyes ëtudiës: (a) RB198, (b) DY50, (c) NBB and (d) MB.

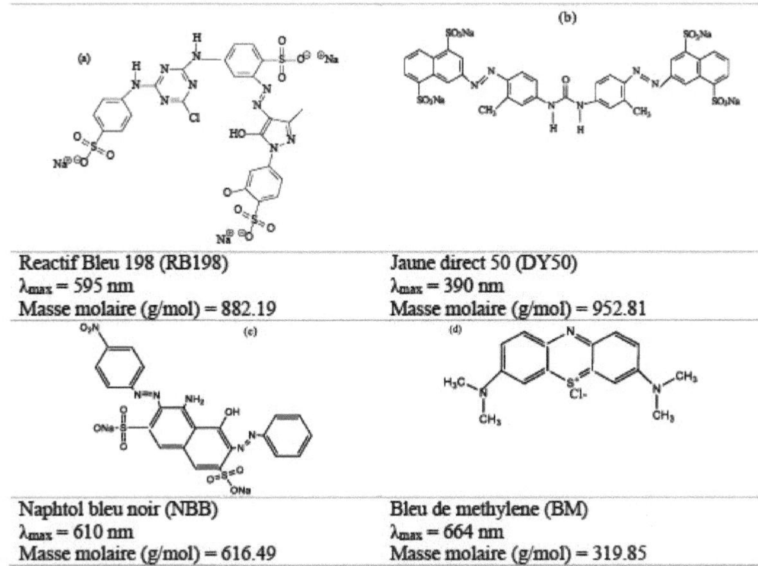

Reactif Bleu 198 (RB198)	Jaune direct 50 (DY50)
λ_{max} = 595 nm	λ_{max} = 390 nm
Masse molaire (g/mol) = 882.19	Masse molaire (g/mol) = 952.81

Naphtol bleu noir (NBB)	Bleu de methylene (BM)
λ_{max} = 610 nm	λ_{max} = 664 nm
Masse molaire (g/mol) = 616.49	Masse molaire (g/mol) = 319.85

RB198 and NBB is shown in Figure II.20a. The adsorbed quantity is maximal at pH = 4 for RB198 and DY50, pH = 8 for BM and pH = 6 for NBB. For example, for BM, the adsorbed quantity increases from 1 to 10.5 mg.g^{-1} when the pH varies from 3 to 9. The lower qt values observed for MB, at acid pH, are due to the presence of excess H+ ions in competition with the BM cations.

Adsorption equilibrium was rapidly reached for dye concentrations ranging from 10 to 30 mg g^{-1} (Figure II.20b-e). Indeed, only 5 minutes of adsorbent-BM contact were sufficient to reach equilibrium, 40 minutes for RB198 and 80 minutes for DY50. This difference in adsorption speed is explained by the reactivity of the adsorbents, the molecular weight and the nature of the dye itself. In addition, functionalization with the copolymer (Figure II.20e) significantly improved the adsorption of NBB. The adsorbed quantity is approximately 5.9 mg.g^{-1} (C0 = 30 mg.L^{-1}) for the optimum dose of copolymer (0.05%). However, it does not exceed 0.25 mg.g^{-1} for the raw membranes under the same conditions. The amount of NBB adsorbed decreases as the dose of cationic agent increases. For example, qt decreases from 5.9 mg g^{-1} to 1.29 mg g^{-1} for the same conditions using a high cationic dose equal to 2%. This result can be explained by the effect of ionic attraction between the cationic groups of the copolymer and the anionic groups of the dye. However, for the modified membranes, the amount adsorbed was slightly reduced at high doses of cationic agent. This is due to the formation of a dye-copolymer complex on the surface of the membranes which, due to steric hindrance, blocks the diffusion of the dye through the hydroxyl groups of the cellulose.

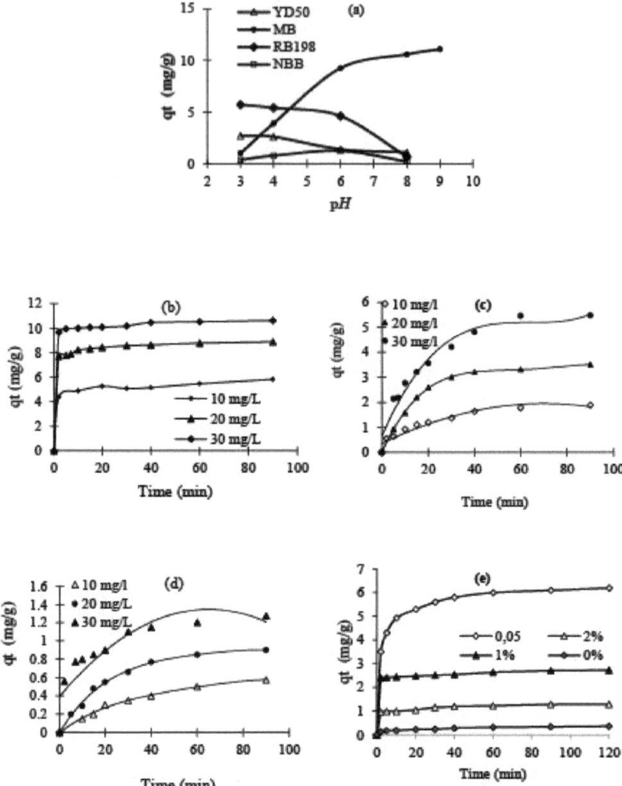

Figure II. 20: (a) Effect of pH (C0= 30 mg/L, t =1h, T=25°C), effect of time: (b) BM, (c) RB198, (d) DY50, and (e) NBB (T= 25°C, pH= 6, C0= 30 mg/L).

The evolution of the adsorbed quantity of BM (Figure II.21a), RB198 (Figure II.21b), DY50 (Figure II.21c) and NBB (Figure II.21d-e) on the surface of the raw and functionalised membranes as a function of temperature indicates that the adsorption process is exothermic. The maximum adsorbed quantities, at 25°C, are, respectively, 16 mg.g^{-1} and 14 mg.g^{-1} for RB198 and DY50.

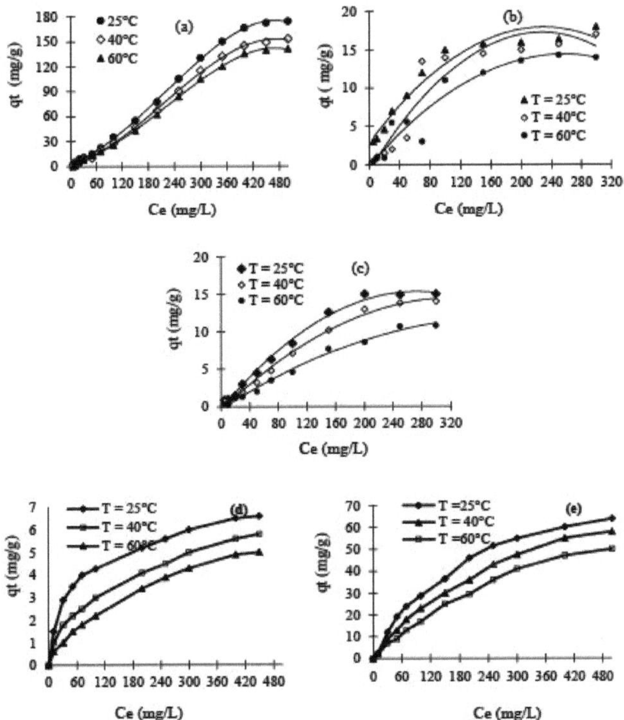

Figure II. 21. Effect of tempërature in the cases of : (a) MB, (b) RB198, (c) DY50, (d) Evolution of tempërature for the NBB dye (case of raw membranes), and (e) case of functionalisedëes membranes.

The adsorbed quantity ëvolves rapidly from 6.6 mg.g^{-1} for the crude membranes (Figure II.21d) to 63.9 mg g^{-1} after functionalisation with the copolymer (Figure II.21e). The quantities adsorbed follow the order: DY50 (14 mg.g^{-1}) <RB198 (16 mg.g^{-1}) <NBB (63.9 mg.g^{-1}) <MB (150 mg.g^{-1}). This difference in adsorption capacity can be explained by the functional groups present in the structure of each dye and also by their molecular weight. DY50 and RB198 have fairly high molar masses (> 882 g.mol)$^{-1}$.

Correlation of the expërimental cinetic data with the cinetic tlreoric ëquations (Table II.13) enabled us to note the following findings; for all the dyes studied, the R2 values turned out to be larger, using the pseudo-second-order equation ($R^2 > 0.85$). The calculated quantities (qe) are also in good agreement with the experimental ones (0.0005 <SSE <0.044). All these findings suggestërent that the adsorption procës is chemical.

Table II. 13: Cinetic constants for adsorption of the dyes ëtudiës on the surface of different membranes

different membranes

c0 (mg /L)	Pseudo-first-order				Pseudo-second order				Elovich			Broadcast	
	K1	qe	R^2	SSE	K2	qe	R^2	SSE	a	B	R^2	K1	R^2
						BM							
10	0.00	2.1	0.66	0.46	0.093	5.7	0.99	0.00	6.65	0.657	0.623	0.40	0.51

T (°C)													
	89	35		1		6		9	4			59	1
20	0.0191	1.8	0.55	0.87	0.003	8.888	0.99	0.002	28.58	0.754	0.529	0.586	0.412
30	0.0098	1.829	0.48	1.1	0.153	10.66	0.99	0.002	303.6	1.119	0.475	0.6697	0.366
RB198													
10	0.007	1.824	0.9	0.006	0.054	2.014	0.97	0.017	0.468	2.493	0.975	0.1988	0.952
20	0.0098	2.9	0.86	0.075	0.099	3.852	0.97	0.044	0.778	1.207	0.971	0.4009	0.904
30	0.0119	4.301	0.92	0.146	0.021	5.875	0.97	0.05	1.656	0.883	0.944	0.56	0.934
DY50													
10	0.0028	1.213	0.94	0.079	0.049	0.724	0.85	0.018	0.068	7.553	0.913	0.0657	0.98
20	0.0038	1.39	0.85	0.06	0.058	1.055	0.94	0.019	0.143	4.57	0.957	0.1042	0.95
30	0.0038	1.376	0.71	0.012	0.14	1.32	0.99	0.0005	0.453	3.731	0.937	0.12	0.83
NBB													
0%	0.003	0.494	0.838	0.015	0.356	0.378	0.983	0.0020	.11	13.5	0.9768	0.036	0.926
0,05%	0.009	2.452	0.801	0.416	0.078	6.305	0.999	0.011	5.2	0.85	0.8279	0.099	0.585
1%	0.005	0.82	0.472	0.212	0.342	2.74	0.999	0.0019	.09	2.504	0.52	0.179	0.416
2%	0.004	0.494	0.621	0.088	0.373	1.309	0.998	0.0021	.84	4.697	0.69	0.529	0.672

Table II.14 shows the constants determined from the Langmuir, Freundlich, Temkin and Dubinin isotherms. The Freundlich isotherm corresponds fairly well to the experimental data with correlation coefficients that are fairly ëlevës ($R2> 0.78$). The cohërence of the Freundlich isotherm with the experimental data reveals that adsorption can occur in heterogeneous adsorption sites. Furthermore, the adsorption of NBB on the surface of raw membranes could be described by the Langmuir equation.

Table II. 14. The Langmuir, Freundlich, Temkin and Redushkevich constants for adsorption of the dyes studied on the surface of cellulose membranes

T (°C)	Langmuir			FreundlichTemkin						Dubinin		
	q_m	R^2	SSE	K_F	n	R B^2		At	R^2	q_m	E R^2	SSE
RB198												
25	16.52	0.81	0.41	0.391	.50	0.781	2.955	0.2	0.946	7.0766	40.0.848	0.447
40	17.85	0.81	0.67	0.	171.29	0.924	2.834	0.13	0.9609	5.2221	19.0.7168	0.47
60	20.242	0.696	1.022	0.2	11.20	0.97	2.443	0.1	0.9092	3.632	28.0.639	0.488
BM												
25	48.309	0.728	8.379	0.021	.14	0.985	4.811	0.12	0.89	5.8814	18.0.5395	11.2
40	51.282	0.78	6.781	0.	001.11	0.9954	.19	0.11	0.896	6.0122	21.0.6431	9.79
60	25.188	0.917	7.72	0.	001.14	0.992	2.564	0.12	0.9	4.1768	35.0.7349	9.12
DY50												

25	48.30 9	0.72 8	2.02	0.	191.14	0.985 4.811		0.1 2	0.895.8814		18.0.53 95	0.8
40	39.21 5	0.63 7	1.614	0. 6	131.13	0.977 3.752		0.1 3	0.885.6933		28.0.65 97	0.7
60	23.80 9	0.73 9	0.92	0.091	.13	0.983 2.329		0.1 5	0.868 4	3.8098	500.73 7	0.41
					NBB : Raw membranes							
25	7.13 2	0.057	0.99 7	0.05 3	1.242	.930 81.35	.	0.929	0.9 8	5.09	25 2	0.87 6	0.15
40	6.70 7	0.029	0.99	0.09	0.592	. 91.3	210.	0.348	0.9 8	4.21	28.86 1	0.80 5	0.158
60	6.21 5	0.02	0.97 9	0.12	0.	311.830 91.24	.	0.256	0.9 7	3.36	40.82 4	0.74 41	0.16
					NBB :Functionalized membranes							
25	84.7 4	0.014	0.87	2.08 5	2.061	. .5	440.815	0.238	0.9 8	38.0 8	17.14	0.92 97	2.58
40	86.2	0.009	0.77	2.82	1.471	.4 814.7	00.	0.171	0.9 8	35	21.32	0.89 5	2.29
60	75.1 8	0.008	0.79	2.52	1.261	. .2	430.812	0.172	0.9 5	26.4	35.35 6	0.70 54	2.35

		Thermodynamic parameters				
		AH* (KJ.mol^{-1}	)AS*(J.mol)$^{-1}$	AG* (KJ.mol)$^{-1}$		
	T (°C)	254060	25 4060	25	40	60
Raw membranes	RB198	-28.255	-132	11.294	13.285	15.939
	DY50	-0.117	-47	14.149	15.825	15.825
	NBB	-24.152	-105	7.273	8.855	10.964
	MB	-0.389	-50	14.744	15.506	16.521
Functionalized membranes	NBB	-12.071	76	10.804	11.956	13.491

The negative enthalpy values suggest that the interaction between ëtudiës dyes and cellulosic membranes is exothermic. This result is in good agreement with both the decrease in adsorption capacity as a function of temperature and the decrease in adsorption energy constants (B) calculated from Temkin liquidation. Positive values of AG* and negative values of AS* indicate, respectively, a non-spontaneous mechanism and a decrease in disorder. In the case of NBB adsorption using functionalised membranes, cationisation increases the disorder of the system.

Conclusion

In summary, this section focused on the exploitation and characterisation of new materials derived from shrimp shells, almond fruit waste, plant fibres and date waste. The functionalization of chitosan with amino silica and the other cellulosic materials with either the chitosan polymer or a copolymer compensated for the low affinity of the cellulosic materials for anionic dyes. All the materials showed very competitive adsorption performances for different classes of dyes: cationic, acid, reactive and direct. In short, in this study we have demonstrated the possibility of recovering new materials from biomass waste, which is very abundant and at no significant cost, for environmental applications in particular.

Ecological production of nanoparticles from biomass for waste treatment applications: Adsorption and degradation of dyes.

Introduction

Currently, the application of nanomaterials has developed in mëdecm [70, 71], sensors [72, 73], water treatment [74, 75] and so on. This means that flora is widely examined. For example, Saranyaadevi et al. described the synthesis of copper oxide nanoparticles using capparious zeylanica extract [76]. Patel et al, reported the synthesis of copper nanoparticles using ocimum sanctum [77]. Angrasan and Subbaiya evaluated the antibacterial activities of copper nanoparticles prepared from vitis vinifera [78]. Nagar and Devra synthesised copper nanoparticles using Azadirachta indica extract [79]. It has been shown that materials enriched in carbonyl, carboxyl, aliphatic hydroxyl and phenolic groups are responsible for the reduction of metal ions to metal oxides. These biomolecules can create a protective environment on the surface of the nanoparticles and contribute to their stabilisation by forming a steric encumbrance around them.

In this context, we propose to synthesise new nanoparticles from lignin extracted from poplar fibres and aqueous extracts of laurel leaves, pergularia leaves and malt fungi. The choice of these extracts is justified by the richness of these biomaterials in active materials capable of reducing metals.

I. Extraction of lignin from poplar fibres: Production of nanoparticles

I.1 Preparation of an alkaline lignin extract from poplar fibres and synthesis of nanoparticles

Firstly, the collected poplar fibres were carefully washed with water to remove sand particles and debris as undesirable materials deposited on the surface. The fibres were then dried in a vacuum oven at 60°C for 24 h and ground into small powders using an original Moulinex grinder. The particles were washed again with distilled water to clean the dust resulting from the grinding process. After cleaning, the fibres were dried at 70°C for 24 h. The dried fibres were then treated with a solution of NaOH (4% w/v) at 80°C for 2 h (1:40 bath ratio) with magnetic stirring. The change in colour of the solution from yellow-white (initial state) to dark black (final state) proves the extraction of the lignin (Figure III.1). The resulting alkaline lignin extract is then used for the production of copper nanoparticles.

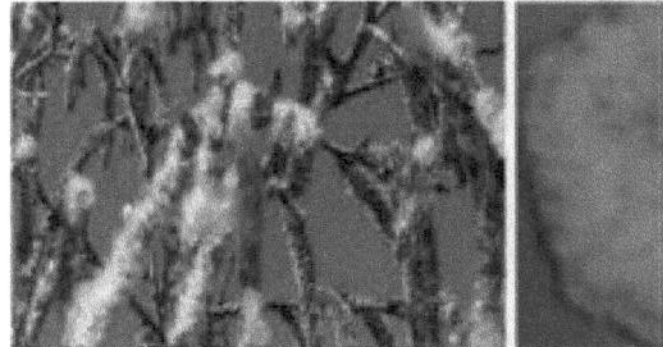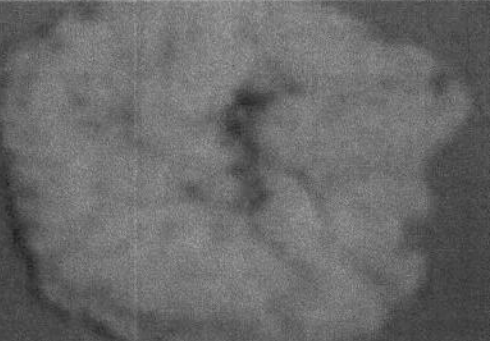

Figure III. 1: Preparation of an alacalin extract of lignin from poplar fibres

The synthesis of copper nanoparticles a ëlë rëalisëe according to the mëthode dëccribed in our previous ëtudes with modifications [80-82]. Here, a 3 mL volume of the freshly prepared alkaline lignin solution obtained from poplar fibres is mixed' with 27 mL of copper sulphate solution (CuSO4 .5H2O, 0.5 M) in a round boutht flask. The mixture was stirred continuously for 60 min at 80°C. The change in colour of the solution during the reaction indicates the reduction of copper sulphate ions to copper nanoparticles. Finally, the solution was centrifuged at 12,000 rpm for 5 min. The precipitation is ^^ë^ as the final product and then analytically analysëe.

I.2 Characterisation of CuO

The formation of copper nanoparticles was first ëlë confirmedëe by observing the change in colour of the copper sulphate solution. In fact, the gradual change in colour from blue to green Амеё a ёlё observed with the addition of alkaline lignin extract to the copper sulphate solution. This proves the reduction of copper ions to copper oxide nanoparticles.

Figure Ш.2 shows the IR-TF spectra of copper sulphate particles, original poplar fibres, extracted lignin and forméed CuO nanoparticles. The CuSO4 spectrum shows absorption peaks at 3094 cm^{-1} , 1652 cm^{-1} , 1063 cm^{-1} and 863 cm^{-1} . The peak at 3094 cm^{-1} corresponds to the hydroxyl group. The peaks observed at lower values correspond to vibrations between the oxygen and non-methyl atoms [83].

The bands around 2916 cm^{-1} and 2852 cm^{-1} are attributed to the CH stretching vibration of the aliphatic groups of the lignin [84]. The bands observed around 1507 cm^{-1} and 1374 -1187 cm^{-1} are respectively attributed to the C=C stretching vibration of the aromatic ring and the C-O stretching vibration of the ester [85]. The bands between 1026 and 718 cm^{-1} are attributed to the aromatic C=O, C-O and C-H vibration groups outside the planar déformation and to the OH groups of the primary alcohol [86].

Briefly, the FT-IR results confirm that the extracted lignin is rich in carbonyl, carboxyl, aliphatic hydroxyl and phenolic groups, which are responsible for the reduction of copper ions to CuO nanoparticles. In fact, these biomolecules can create a protective environment on the surface of the nanoparticles and contribute to their stabilisation by forming a steric packing around them [87, 88]. This trend suggests that the alkaline extract of the studied lignin acts as a reducing agent and a support for the formed nanoparticles.

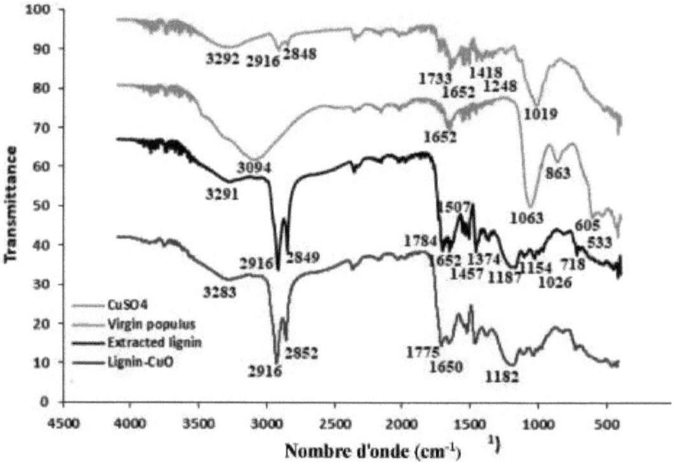

Figure III. 2. IR-TF spectra of copper sulphate particles, poplar fibres, extracted lignin and CuO

lignin and CuO formate nanoparticles

Diagram III.1 shows a mechanical route for the production of copper oxide nanoparticles. Firstly, the lignin could be dissolved in an alkaline solution in which CuOH could be formed. Secondly, the aliphatic hydroxyl groups of the lignin could form complexing agents with the copper sulphate as precursor salts and thus form a ligand with the copper ions. This starts the nucleation process, which then leads to the formation of metal oxides.

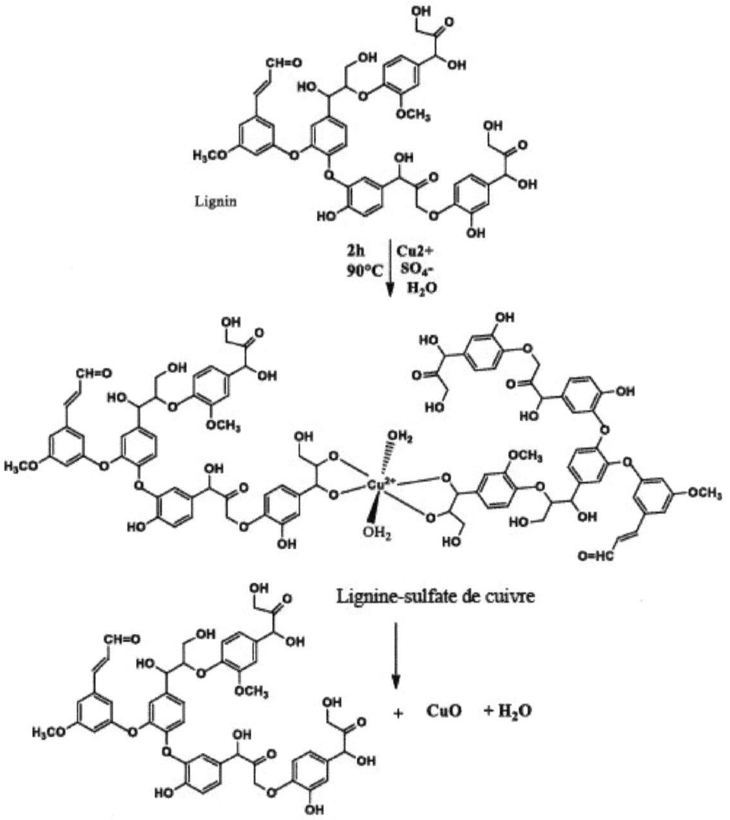

Schema III.1: Probable mechanism for the formation of CuO nanoparticles from extracted lignin.

extracted lignin

Figure III.3 shows SEM images, rëalisëes at more ëlevës magnifications (x750, x 1500 and x3000), of the prepared copper oxide nanoparticles. From these images, it can be seen that the synthëtisëed nanoparticles are sphërical in shape. Some nanoparticles appeared quite sëparëes from each other and some of them are agglomërëes due to oxidation of the mëtalic nanoparticles.

Figure III. 3: SEM images of nanoparticles formed from the alkaline extract of lignin

The EDX spectrum, given in Figure III.4, shows an intense peak at 0.9 and less intense peaks at 8.1 and 8.9 keV representing CuLa, CuKa and СиКв [89], contributing respectively to a weight percentage of 43.69%. An intense peak of elemental oxygen was observed at 0.5 keV with a weight percentage of 39.84%, confirming the formation of CuO. The presence of S at 2.4 KeV with a weight percentage of 16.47% indicates residues from the oxidation process [90, 91].

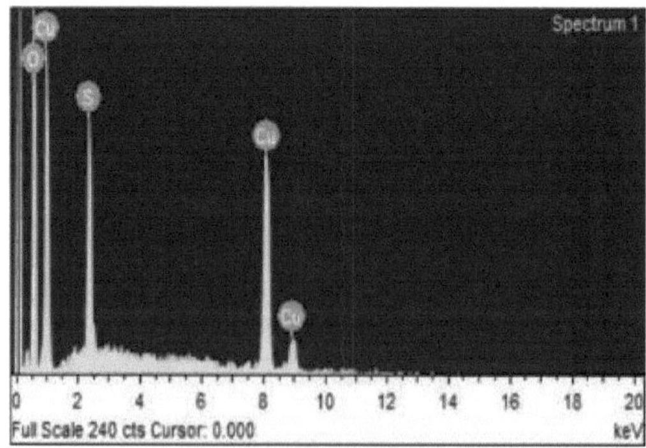

Elements	Weight	Atomic
O K	39.84	67.46
S K	16.47	13.91

| Cu L | 43.69 | 18.63 |
| **Total** | 100.00 | |

Figure **III. 4** EDX scans and chemical composition of nanoparticles formed from the alkaline extract of lignin.

alkaline extract of lignin

The XRD spectrum of CuO nanoparticles is shown in Figure III. 5. The spectrum shows values from 26 to 33.6°, 35.9, 38, 46.5, 50.5, 59.4, 62, 65.5, 67.2, 72.5 which correspond to the planes (1 1 0), (1 1 -1), (1 1 1), (2 0 -2), (1 1 2), (2 0 2), (1 1 -3), (3 1 0), (1 1 3), (2 2 1) planes of the copper oxide (CuO) monoclinic phase (JCPDS No : 98-009-2367). The presence of diffraction peaks at around 26 = 35-39° proves the formation of CuO [92]. The peaks observed at 30.8° and 41.6° correspond to the (0 1 1), (0 0 2) planes of Cu2O (JCPDS No: 96-9005770). The peak observed at 43.5° is attributed to the (1 1 1) Cu plane (JCPDS No: 04-0836). The other peaks observed between 10° and 30° reveal the presence of reducing groups on the surface of the copper nanoparticles. The results obtained are in agreement with the literature reporting the preparation of copper oxide nanoparticles from other biological extracts [93-95].

Using Scherrer's equation, the average crystallite size of copper oxide nanoparticles was determined to be 12.4 nm.

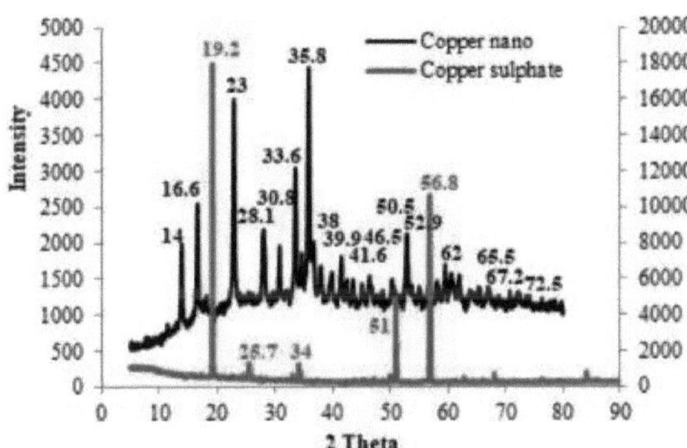

Figure **III. 5:** XRD spectrum of copper sulphate and nanoparticles formed from the alkaline extract of lignin.

alkaline extract of lignin

TEM images showed a sphërical morphology (Figure Ш.6). Overall, the nanoparticles presented a sphërical shape structure. The overlapping of the nanoparticles results in different patterns justified by the presence of the alkaline lignin extract on the surface of the CuO particles. TEM results are also consistent with IR-TF, SEM and XRD results.

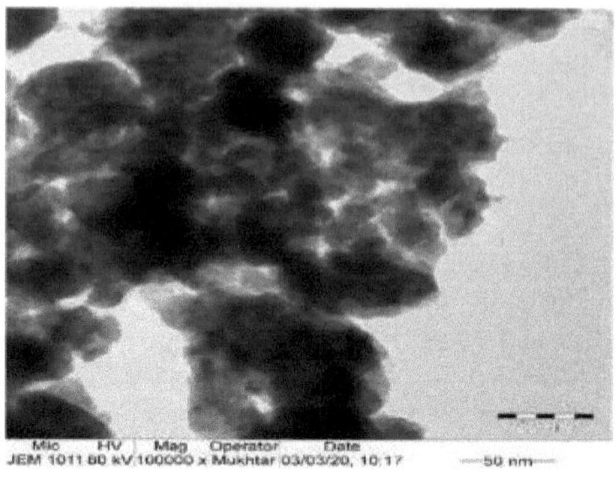

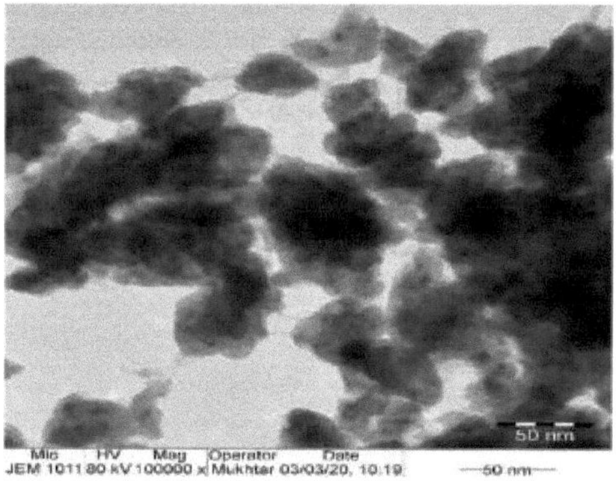

Figure III. 6 TEM images of nanoparticles formed from the alkaline extract of lignin

I.3 Application of nanoparticles prepared for methylene blue adsorption

1.3.1 Influence of experimental parameters on methylene blue adsorption

The initial pH of the solution, the duration of the adsorbent-adsorbate contact, the concentration of methylene blue, and the tempërature of the solution were etë ëtudiës in order to define the optimum conditions for the biosorption process in the presence of CuO nanoparticles.

The data in Figure III.7a show that the biosorption of methylene blue increases rapidly as the initial pH value of the coloured solution rises from 3 to 6, above which no further adsorption is observed. These results could be interpreted by the fact that under strongly acidic pH conditions, CuO nanoparticles are positively charged due to protonation. In fact, this protonation opposes the positively charged methylene blue ions, thus lowering the adsorption capacity under these conditions. The optimum pH value to achieve maximum adsorption is observed at value 6. This suggests that the surface of the nanoparticles becomes negatively charged due to deprotonation of the oxygen groups leading to electrostatic interaction with the cationic dye molecules.

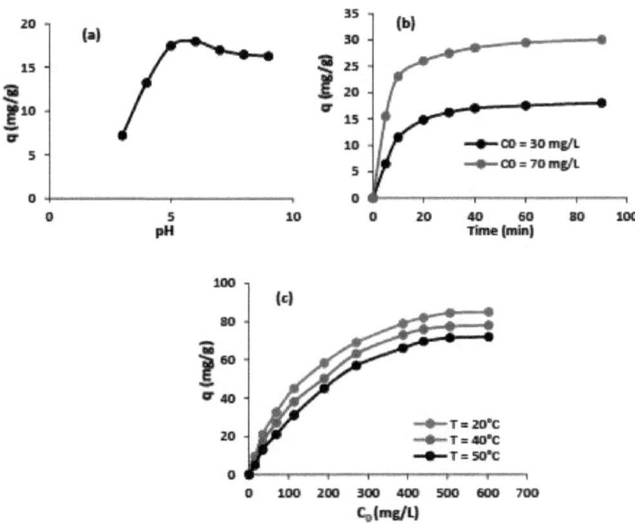

Figure III. 7: Variation of adsorbed capacity as a function of : (a) pH (T= 20°C, time = 60 min), (b) time (T= 20°C, pH = 6), and (c) temperature (pH = 6, time = 60 min).

The effect of reaction time on the biosorption of metelia blue using CuO nanoparticles is shown in Figure III.7b. The adsorption capacity increases rapidly during the first 20 min and reaches equilibrium at 40 min of contact. Indeed, more than 85% of the target was achieved within the first 20 min. This result demonstrates the effectiveness of using CuO nanoparticles to remove metiylene blue from water. In fact, numerous unoccupied active sites are available on the surface of the adsorbent studied during the first period of time, which are responsible for the large adsorption of the dye molecules. After 40 minutes of reaction, the surface of the adsorbent is partially filled and consequently no further increase in adsorption capacity has occurred.

According to the data provided in Figure III.7c, the adsorption of metiylene blue is rapid at low dye concentrations and becomes almost stable when the concentration reaches high values. The high adsorption capacities at lower dye concentrations are due to the availability of adsorption sites on the surface of the CuO nanoparticles. When equilibrium is reached, the maximum adsorption capacity is 85 mg/g (pH = 6, t = 60 min, T = 20°C). This adsorption capacity is significant. The decrease in adsorption capacity with increasing temperature can be attributed to the escape of adsorbed metiylene blue ions at a higher energy or temperature. Adsorption follows an exothermic mode. The adsorption capacity decreases from 85 mg/g to 72 mg/g when the temperature is increased from 20°C to 50°C. Compared with biosorbents previously reported for metiylene blue and under the same conditions, the adsorption capacity obtained for CuO nanoparticles (85 mg/g) is, for example, much greater than jute deciets (22.47 mg/g) [96], and orange peel (18.6 mg/g) [97]. It is comparable to sodium alginate gel beads modified with a deporpiyrin-zinc(II) complex (52.3 mg/g) [98], copper oxide prepared from malt wax extract (64 mg/g) [81], and copper oxide prepared from rose laurel (81.2 mg/g) [80].

I.3.2. Kinetic study

At equilibrium, an understanding of the interaction between methylene blue molecules and CuO nanoparticles could be interpreted by modelling the experimental kinetic data using the pseudo-first-order, pseudo-second-order, Elovich and intra-particle diffusion equations. This modelling provides information on the biosorption mechanism if it is a physical, chemical and/or mass transfer phenomenon. The values of the correlation coefficients (R^2) for the pseudo-second order kinetic equation are equal to 0.99 which are superior to the other kinetic equations studied (Figure III.8, Table III.1). In addition, the calculated adsorption capacity values, determined for the pseudo-second order model, showed excellent agreement with the experimental values. These trends indicate the chemical nature of the methylene blue adsorption process on the surface of CuO nanoparticles. The relationship between qt and $t^{1/2}$ for the intra-particle diffusion equation (Figure III. 8d) shows a clear divergence from the origin of the graph. This confirms that this cinetic model is not the only step controlling the rate, but that other cinetic processes could exist during adsorption.

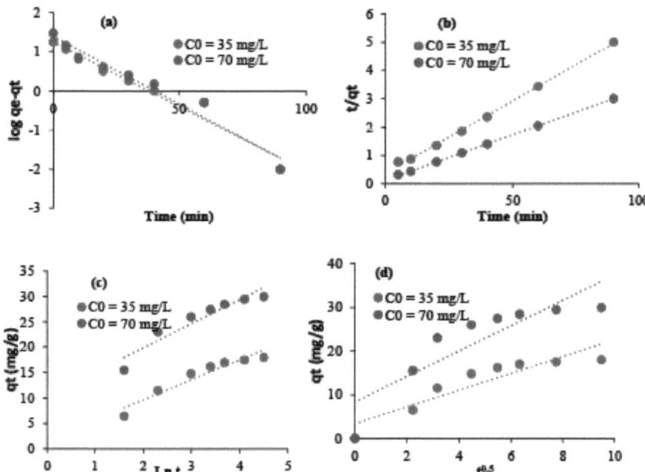

Figure III. 8: Cinetic modelling: (a) pseudo first-order, (b) pseudo second-order, (c) Elovich, and (d) intra-particle diffusion.

Table III. 1: Kinetic constants, isotherms and thermodynamic parameters related to the adsorption of methylene blue in the presence of CuO nanoparticles.

Pseudo cinetic equations	Constants	Concentration (mg/L) 30	70	Isotherms	Parameters	Temperature (°C) 20	40	55
first	K_i (min)$^{-1}$	0.033	0.035		q_m (mg.g^{-1})	105.26	104.16	103.09
order	q (mg.g^{-1})	17.78	24.95	Langmuir	K_L (L. g)$^{-1}$	0.0074	0.0058	0.0042
	R^2	0.96	0.96		R^2	0.99	0.99	0.98
Username second	K_2	0.0067	0.0072	Parameters thermodynamics	ДH° (KJ mol)$^{-1}$		-4.17	
order	q	19.72	31.55		AS° (J mol)$^{-1}$		-30.49	
	R^2	0.99	0.99		AG° (KJ mol)$^{-1}$	8.93	9.54	9.85
					K_{pf} (L.g)$^{-1}$	13.17	4.95	1.54

Elovich	a(mg.g^{-1}.min)$^{-1}$	9.26	27.11	Freundlich	n	1.83	1.66	1.49
	$\textit{б}$ (mg.g^{-1}.min)$^{-1}$	0.25	0.21		R^2	0.98	0.98	0.98
	R^2	0.92	0.90		b_T (J.mol)$^{-1}$	117.31	131.34	142.33
Broadcast	K(mg.g^1.min /)^{12}	1.92	2.92	Temkin	A(L.g)$^{-1}$	9.91	11.71	13.86
intra-	R^2	0.84	0.77		R^2	0.96	0.95	0.94
particulate								

I.3.2. Adsorption isotherms

The isotherms could clarify the behaviour of the interaction between the adsorbent and the adsorbate as well as the distribution of the adsorbate between the solid and the solution phase during the adsorption process. In this study, the isotherms were studied using the Langmuir, Freundlichet Temkin models. The graphs, corresponding parameters and correlation coefficients are given in Figure III.9 and Table III.1, respectively.

The calculated model parameters show better agreement with the Langmuir and Feundlich models ($R2 > 0.98$). This suggests that adsorption occurs in monolayers and multilayers. The decrease in KF values with increasing temperature values from 13.17 to 1.54 proves that the lowest adsorption capacities occur at higher temperature values. The BT values calculated from Temkin's equation, related to heat adsorption, are between 20.76, 19.81 and 18.87 J/mol, respectively, for temperatures 20, 40 and 55°C. These values are greater than 1, suggesting electrostatic interaction and pore heterogeneity on the biosorbent surface. The decrease in these constants with increasing temperature is consistent with the decrease in methylene blue adsorption capacity with increasing temperature.

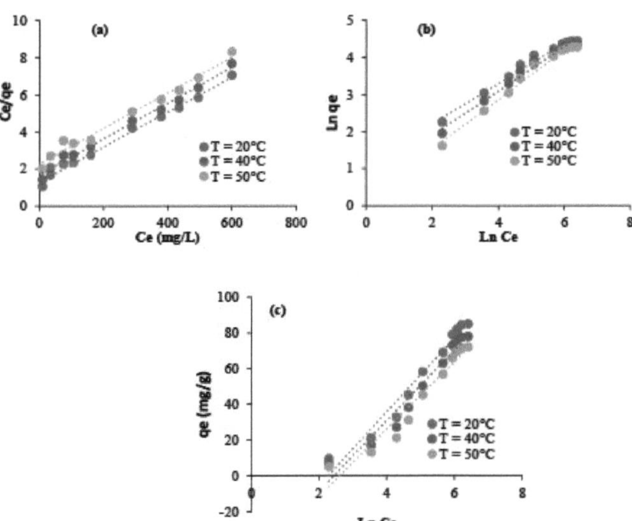

Figure III. 9. isotherm plots for 1 adsorption of тёЛу1епе blue in the presence of тёЛу1епе blue.

CuO nanoparticles: (a) Langmuir, (b) Freundlich and (c) Temkin

The thermodynamic parameters (AH° and AS°) have ёle calculated from the data obtained from the №ace of Ln Kd as a function of 1 inverse of tempёrature (Figure Ш.10). The negative enthalpy value (AH° = -4.17 Kj/mol) confirms that the phёnomёne of adsorption of тёЛу^ис

71

blue to the nanoparticle surface is exothermic, which is in agreement with the results of the tempёrature effect. The calculated positive Gibbs energy values (AG° = 8.93 - 9.85 Kj/mol) show a non-spontaiKd process.The ^gative entropy value (AS° = -30.49 j/mol) rёyёк the decrease in the ateatory character of the adsorbent-solution interface upon adsorption of mёthylёne blue.

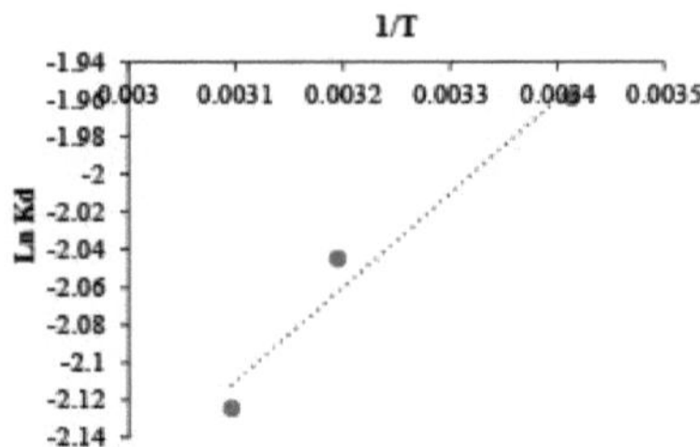

Figure III. 10. plot of Ln Kd versus 1/T Hё on adsorption of тёШy1епе in the presence of prepared CuO nanoparticles.

II. Ecological synthesis of nanoparticles from laurel leaves, pergularia leaves and malta fungi, characterisation and application

II.1 Synthesis of nanoparticles from laurel leaves, pergularia leaves and malta mushrooms

The pergularia leaves were grown in the Sidi Bouzid-Tunisia region during the winter (January). Laurel leaves and malta mushrooms were collected in the Skanes-Monastir region. The collected biomasses were washed with distilled water to remove any impurities deposited on the surface and then dried in the dark for approximately 10 days. A ёelectric mёlangeur a ёlё utilisё to grind the biomass fractions into fine powders. 10 g of the ground fractions were mixed with 150 mL of distilled water and heated in a flask for 90 minutes at 70°C. Afterwards, the coloured solution was cooled to room temperature and filtered. Next, the coloured extract obtained (100 mL) was added to 100 mL of a copper sulphate solution ($CuSO_4.5H_2O$, 1M). The mёtalic salt solution and biological extract were ёlё agile's at 70°C for 120 minutes. After a certain time, the colour changed, indicating the reduction of the copper salt to copper oxide nanoparticles (Figure III.11). At the end of the reaction, the mixture was centrifuged for 20 minutes. The nanoparticles obtained were ёtё sёchёes at 150°C for 12 hours.

Figure III. 11. Photographs showing the different ëtages of synthesis of copper oxide nanoparticles
oxide nanoparticles from pergularia leaves, laurel leaves and malta mushrooms.

II.2. Adsorption and catalytic reduction tests on methylene blue

Adsorption experiments were carried out in batch mode using Erlenmeyer flasks containing 0.0125 g of synthesised nanoparticles and 20 mL of methylene blue, with magnetic stirring at 125 rpm. After each experiment, the liquid was filtered using filter paper and its absorbance was evaluated using a spectrophotometer.

Catalytic reduction of methylene blue was carried out in the presence of NaBH4 as the reducing agent. A volume of 8.5 mL of NaBH4 solution (0.67 g/L) was mixed with 1.5 mL of methylene blue solution (0.32 g/L). To this solution (volume = 10 mL), a quantity of nanoparticles (0.005 g) was added. The solution was stirred at regular time intervals at room temperature and analysed at the maximum wavelength 665 nm. The decolourisation yield was calculated using the following formula:

$$R(\%) = \frac{(A0 - At)}{A0} \times 100$$

Where A0 and $_{At}$ are the absorbances measured at time t = 0 and time t, respectively.

To better understand the kinetics of catalytic reduction of the methylene blue solution and to determine the pseudo-first order rate constant, $_{ko}$, the values of Ln Ct/Co were plotted as a function of time:

$$\ln C_t/C_0 = -k_0 t$$

Where t is the time studied during degradation, k0 is the first-order rate constant and C and C0 are defined as the concentrations of methylene blue at times t and 0.

II.3 Characterisation of synthesised nanoparticles

The IR spectra of pergularia leaves, laurel leaves, malted mushrooms, and copper oxide nanoparticles synthesised from their aqueous extracts are shown in Figure III.12. For the spectra of the biomasses studied, the bands recorded at 3312-3199 cm^{-1} , 2799-2915 cm^{-1} are attributed, respectively, to hydroxyl (OH) and CH groups of the aliphatic compounds [99]. The bands observed at 1711-1721 cm^{-1} , 1544 cm^{-1} and 1378-1211 cm^{-1} prove, respectively, the presence of C = O, C = C, and C-O groups [99]. The absorption band observed at 993-1104 cm^{-1} corresponds to the C-O-C group [100]. The particularity of the IR spectra of copper oxide nanoparticles compared with those describing the starting biomasses lies in the fact that the main peaks characteristic of the nanoparticles have undergone chemical displacements. For example, the band relating to the OH group (3312 cm^{-1}) in the case of pergularia leaves shifted to a higher value (3510 cm^{-1}). The ester group band shifted to 1607 cm^{-1} . These findings suggest that the molecules in the biomass extracts reacted with the copper ions. A search of the literature has shown that these biomolecules are responsible for reducing and protecting nanoparticles against oxidation [101-104].

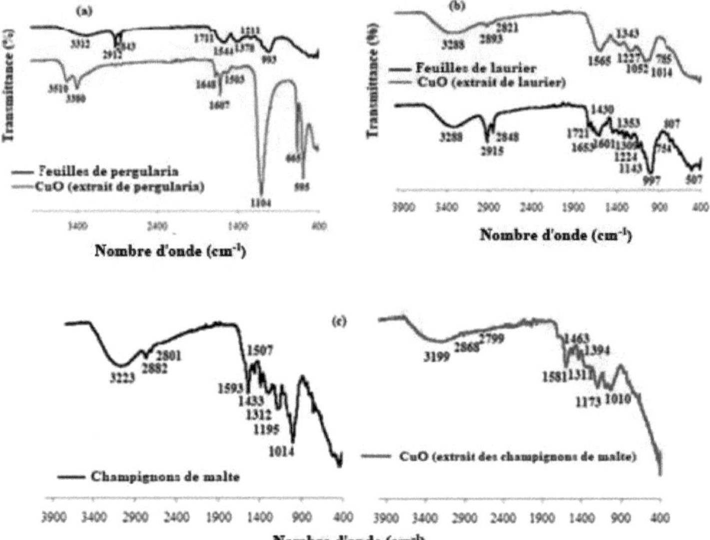

Figure III. 12. IR spectrum of pergularia leaves, laurel leaves, malta mushrooms, and nanoparticles synthëtisëed from their aqueous extracts.

SEM photographs of copper oxide nanoparticles synthesised from extracts of pergularia leaves, bay leaves and malta mushrooms are shown in Figures III.13-15. The particles appear spherical with some agglomeration. This agglomeration could be due to the hydroxyl groups present in the biological extracts studied. This difference in shape distribution could be attributed to the difference in composition of the extract itself. Chemical analysis of the nanoparticles by EDX (Figure III.13c, Figure III.14c, and Figure III.15c) shows the presence of a characteristic peak around 1 Kev which is a CuO index. The signals corresponding to the carbon and oxygen atoms indicate the presence of photochemical compounds in the biological extracts. The low quantities of P (2.26Z), N (5.83Z), Al (1.24Z) and Fe (0.36Z) atoms observed in the case of nanoparticles prepared from pergularia leaf extract are characteristic of the starting product.

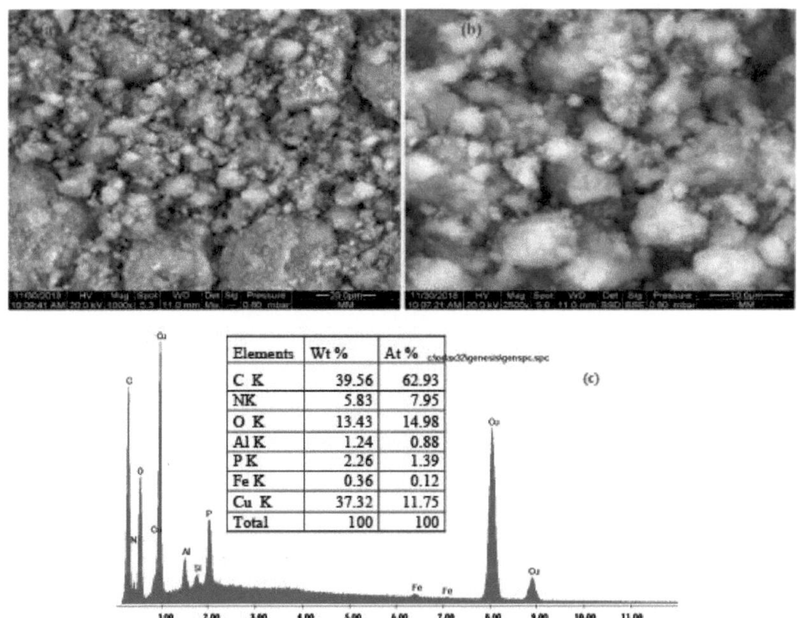

Figure III. 13. SEM photos of nanoparticles prepared from pergularia leaves observed at different magnifications: (a) *1000, (b) x 2500, and (c) EDX analysis.

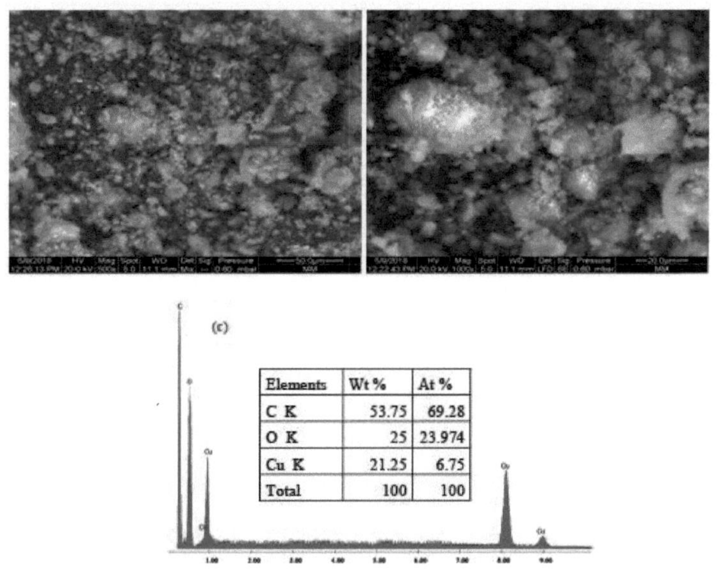

Figure III. 14. SEM photos of nanoparticles prepared from laurel leaves observed at different magnifications: (a) x500, (b) x1000, and (c) EDX analysis.

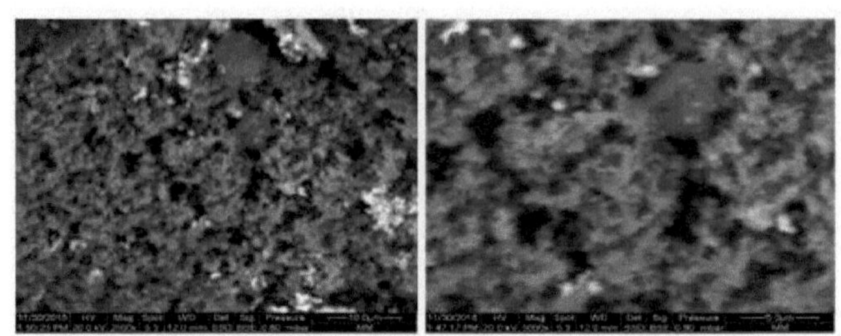

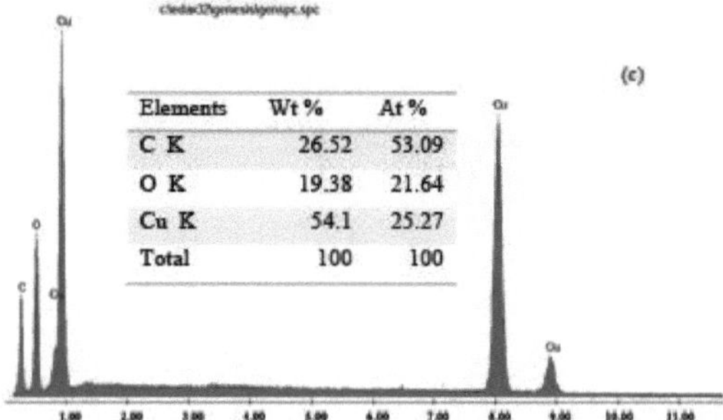

Elements	Wt %	At %
C K	26.52	53.09
O K	19.38	21.64
Cu K	54.1	25.27
Total	100	100

(c)

Figure III. 15. SEM photos of nanoparticles prepared from malt fungi
observed at different magnifications: (a) *2500, (b) *5000, and (c) EDX analysis.

The mechanism by which CuO nanoparticles are formed is based on the complexation of biomolecules from biological extracts of pergularia leaves, bay leaves and malt fungi with copper ions. In fact, the biomolecules first form ligands with the Cu ions^{2+} through OH. The nucleation process then reduces the Cu^{2+} ions to nanoparticles. At high temperatures (>80°C), the copper-ligand complex decomposes easily, releasing CuO [105]. Quercetin malonylhexoside and isorhamnetin 3-O-malonylhexoside are the main constituents of pergularia tomentosa leaves [106]. Cinnamic acid is the main compound identified in bay leaves. Catechin and chlorogenic acid are secondary constituents [107]. The main constituents of malt fungi are phenolic compounds, steroids, and triterpenes [108]. Scheme Ш.1 predicts the formation mechanism of copper oxide nanoparticles using biological extracts of pergularia leaves, laurel and malta mushrooms.

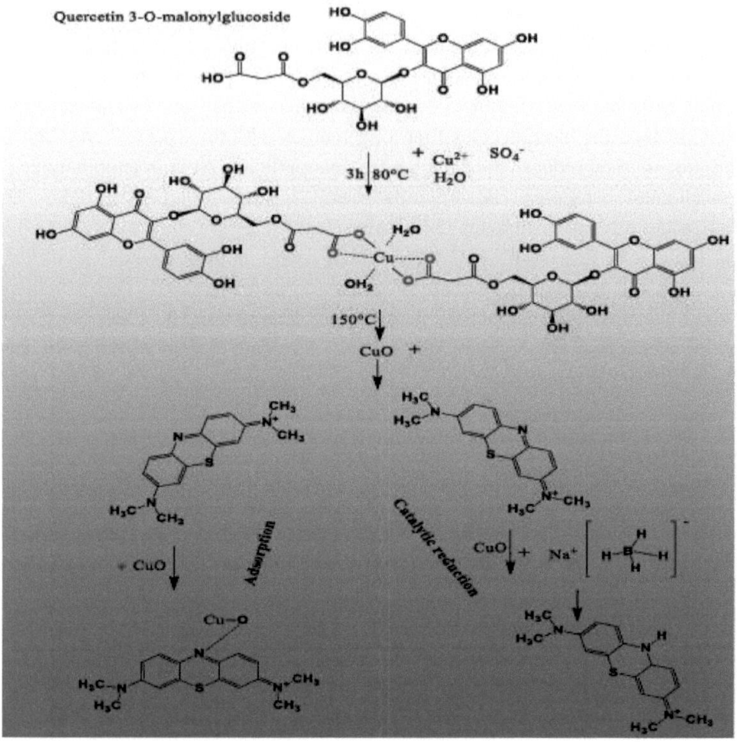

Quercetin 3-O-malonylglucoside

3h 80°C
H_2O

$+ \ Cu^{2+} \quad SO_4^-$

150°C

CuO +

Adsorption

Catalytic reduction

$+ CuO$

$CuO + Na^+$

Cu—O

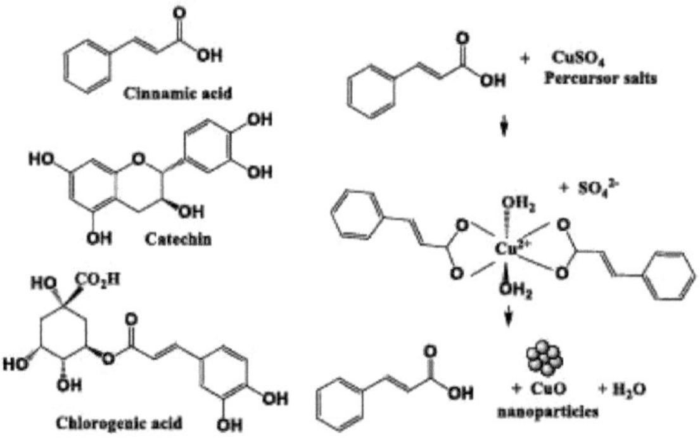

Cinnamic acid

Catechin

Chlorogenic acid

$+ \ CuSO_4$
Percursor salts

$+ \ SO_4^{2-}$

$+ \ CuO \quad + H_2O$
nanoparticles

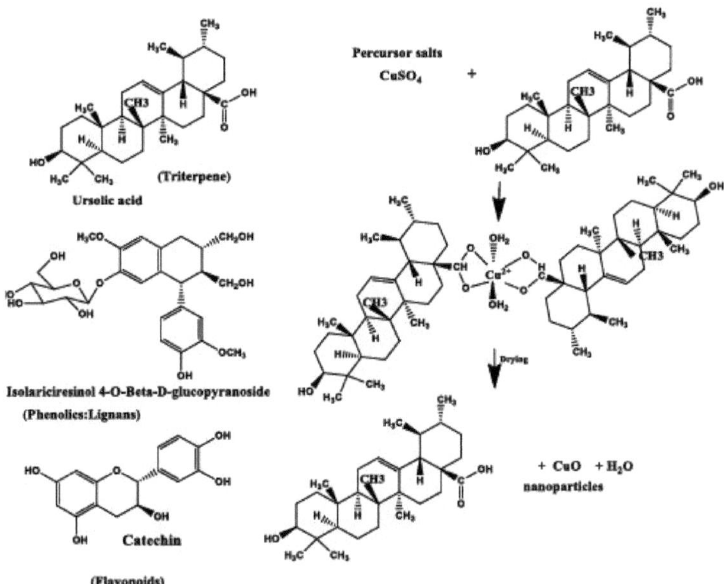

Diagram III.1. Probable mechanism dëcribing the formation of copper oxide nanoparticles copper oxide nanoparticles: (**a**) prepared from pergularia leaves, (**b**) laurel leaves, and (**c**) malta
mushrooms.

The DRX analyses of copper oxide nanoparticles prepared from biomass ëtudiëes have been reprteësentës in Figure Ш.16. In the case of nanoparticles prepared from pergularia leaves, a sërie of peaks was recorded at 29 = 25.4, 31.2, 42.15, 51.84, 62.8 and 76.3. Indeed, the peaks recorded at 29 = 42.15, 51.84.5 and 76.3 were identifiedës from the (111), (200) and (220) planes. The same findings were observed for nanoparticles prepared from bay leaves and malta mushrooms. These data suggest that the synthesised nanoparticles have a face-centred cubic structure (JCPDS No. 85-1326). The results obtained are in good agreement with those reported in the literature [109].

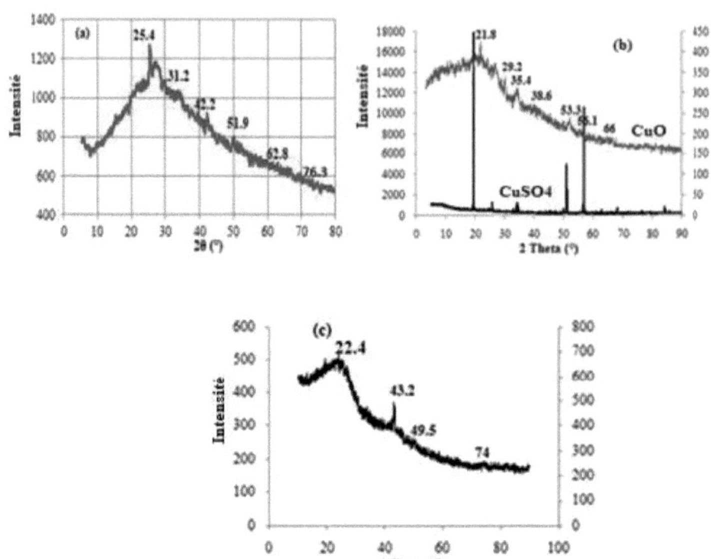

Figure III. 16. XRD analysis of copper oxide nanoparticles prepared from:
(a) pergularia leaves, (b) laurel leaves, and (c) malt mushrooms.

The morphological characteristics of copper oxide nanoparticles synthëtisëed from pergularia leaves were ële ëvaluatedë by TEM (Figure Ш.17). The nanoparticles are sphërical with sizes ranging from 1.7 nm to 15.1 nm. This size diffërence suggëre the difference in the chemical composition of the extract. These data areëë also in agreement with the SEM analyses.

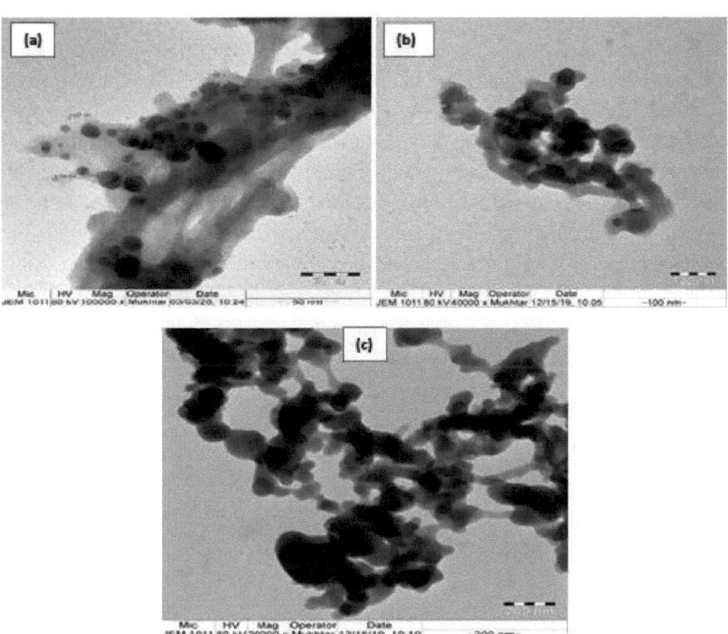

Figure III. 17. TEM images of nanoparticles prepared from pergularia leaves:
(a) x 100000, (b) x 40000, and (c) x 20000.

II.4. Application of nanoparticles for the adsorption of methylene blue

The influence of varying the initial pH value on the adsorption of тёЛу1еne blue using copper oxide nanoparticles prepared from pergularia leaves, bay leaves, and malt mushrooms is shown in Figure III.18a, Figure III.19a and Figure III.20a. We observe that the quantity of methylene blue adsorbed on the surface of the nanoparticles increases with increasing pH value and reaches its maximum at pH = 6. The adsorbed quantity of methylene blue at pH values below 6 could be interpreted by the fact that many H+ protons are available on the surface of the adsorbent. This leads to electrostatic repulsion forces between methylene blue as a cationic dye and the adsorbed H^+ ions. At higher pH values and more precisely when pH = 6, the improvement in adsorption capacity could be justified: by ёlectrostatic forces of attraction between the cationic dye and the negatively charged surface of the nanoparticles.

Figure III.18b, Figure III.19b and Figure III.20b show the involution of the adsorbed quantity of methylene blue as a function of time. In the case of nanoparticles prepared from pergularia leaves, methylene blue adsorption equilibrium is reached after 40 minutes of contact. In fact, more than 90% of the target was reached in the first 20 minutes. This translates into the fact that during this first ёst stage, numerous active adsorption sites are available on the surface of the nanoparticles. The adsorption rate slowed down after 20 minutes of reaction, as the active sites became more saturated. After 40 minutes of reaction, no further adsorption took place. In the case of nanoparticles prepared from bay leaves and malt fungi, equilibrium at ёle approximately reached after 30 minutes of contact. This rapidity of reaching adsorption equilibrium proves the efficiency of using these nanoparticles as adsorbents of cationic dyes.

Figure III.18c, Figure III.19c and Figure III.20c plot the adsorbed quantity of methylene blue as a function of dye concentration for nanoparticles prepared from biomasses. The maximum adsorbed quantity is 93.20 mg/g at 22°C in the case of nanoparticles prepared from pergularia leaves. It is equal to 81.2 mg/g at 22°C in the case of nanoparticles prepared from laurel leaves. It is 64 mg/g at 20°C in the case of nanoparticles prepared from malt mushrooms. We also note that these quantities depend on temperature and they adopt an exothermic behaviour in the case of nanoparticles prepared from pergularia leaves and an endothermic behaviour in the case of nanoparticles prepared from laurel leaves and malt fungi. For example, in the case of nanoparticles prepared from pergularia leaves, by increasing the temperature from 22°C to 55°C, the quantity of adsorbed dye decreases from 93.2 to 83 mg/g (Figure III.18c). These recorded adsorption capacities are interesting and therefore nanoparticles prepared using biological extracts could be considered as good adsorbents of cationic dyes.

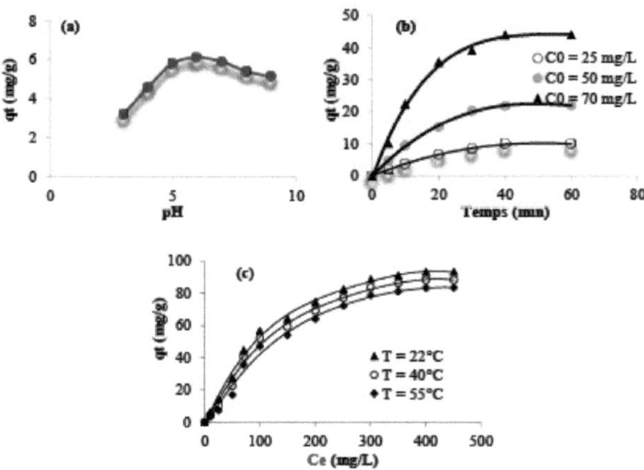

Figure III. 18. (a) Effect of pH (C0 = 10 mg / L, T= 22 °C, time = 40 min), (b) effect of reaction time, (c) effect of dye concentration and tempërature on the capackë adsorption in the presence of nanoparticles prepared from leaves of pergularia

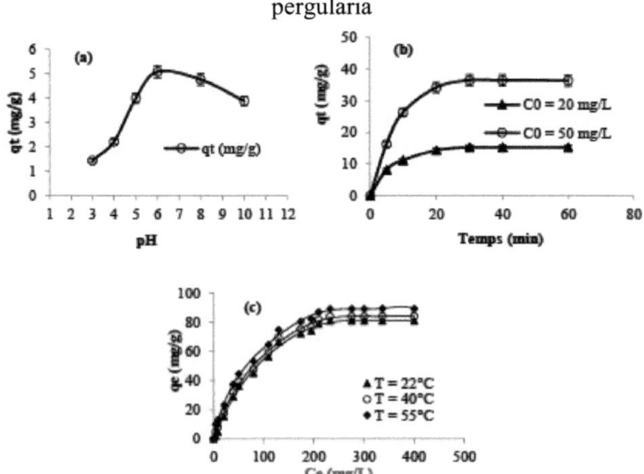

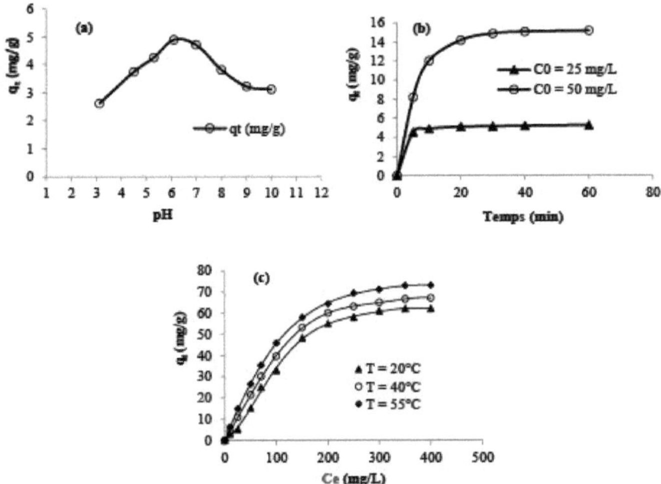

Figure III. 19. (a) Effect of pH (C0 = 10 mg/L, T= 22°C, time = 40 min), (b) effect of reaction time
, (d) effect of dye concentration and tempërature on the capackë
adsorption in the presence of nanoparticles prepared from Laurier leaves.

Figure III. 20. (a) Effect of pH (C0 = 10 mg/L, T= 22°C, time = 40 min), (b) effect of reaction time
, (d) effect of dye concentration and tempërature on adsorption capacity in the presence of nanoparticles prepared from malt fungi.

adsorption capacity in the presence of nanoparticles prepared from malt fungi

The understanding of the mëcanism of тёЛу^ис blue adsorption using synthëtisëed copper oxide nanoparticles has ëlë ëtudiëe by using rëfërant to pseudo-first-order, pseudo-second-order, Elovich and intra-particle diffusion ciiretic equations. The ciretic paramëtres calculated from the modëles ëtudiës have ëtë rësumësed in Tables III.2-4. We note that the correlation coefficient recorded for pseudo second-order ciretic liquidation $(0.95 < R^2)$ are larger compared to pseudo first-order ones $(0.81 < R^2 < 0.95)$ for nanoparticles prepared from the three biomasses. This result suggestsëre that the mëthylëne blue modules have ëlë adsorbed chemically. Tracës of the ciretical intra-particle diffusion modële also showed fairly reasonable correlation coefficients $(R^2 > 0.93)$ suggërant son applicabil^ pour dëcrire l'adsorption du bleu de mëthylëne dans le cas des nanoparticules prëparëes a partir des feuilles de pergularia. This implies, in this case, that the adsorption rate is controlled by intra-particle diffusion. The correlation coefficients for Elovich liquation are large $(0.96 < R^2)$ in the case of nanoparticles prepared from pergularia leaves, which means that the adsorption mechanism is chemical with heterogeneous pores on the surface of the adsorbent. Based on all these findings, the adsorption mechanism remains complex to understand and many steps may be involved.

Table III. 2. Summary of kinetic constants, Langmuir,
Freundlich, Temkin, and Dubinin parameters

for methylene blue adsorption in the presence of nanoparticles prepared from pergulariaCuO

Kinetic equations	Constants	Dye concentration (mg/L)			Isotherms	Parameters	Temperature (°C)		
First username		25	50	70			22	40	55
order	K_i (min^{-})^1	0.037	0.043	0.048		$q_m^{(m}$ g.g$^{-1)}$	136.99	151.52	172.4
	$q_C$mg.g)$^{-1}$	12.79	28.84	59.29	Langmuir	K_L(L. g)$^{-1}$	0.0056	0.00376	0.00254
	R^2	0.93	0.95	0.93		R^2	0.97	0.89	0.73
Pseudo-second	K_2	0.0022	0.0013	0.0009	Thermodynamic	$\Delta H°$ (KJ mol^{-1}		-19	
Order	q	15.77	32.79	60.2	parameters	) $\Delta S°$ (J mol^{-1}		-107.43	
	h	0.543	1.37	3.31		$\Delta G°$ (KJ mol^{-1}	12.692	14.626	16.237
Elovich	R^2 o(mg.g^{-1}	0.96	0.95	0.96	Freundlich	K_F(L.g)$^{-1}$	1.628	0.935	0.609
	.min^{-1})	1.209	2.91	6.77		n	1.419	1.26	1.168
	β (mg.g^{-1} .min^{-1})	0.284	0.131	0.07		R^2	0.95	0.94	0.94
	R^2	0.98	0.97	0.96	Temkin	B_T (J/mol)	26.071	25.397	24.526
Broadcast	K(mg.g^{-1} .min$^{1/2}$)	1.42	3.21	6.58		A_T(L.g)$^{-1}$	0.0852	0.0756	0.0861
Intra-						bt (J/mol)	92.913	102.46	111.19
particulate	R^2	0.96	0.95	0.93		R^2	0.97	0.97	0.96
					Dubinin	q (mg g$^{-1)}$	63.225	57.57	51.357
						E (KJ/mol)	100	100	111.8
						K_{DR} (mol /kJ)^{22}	5x10^{-5}	5x10^{-5}	4 x10^{-5}
						R^2	0.70	0.70	0.65

Table III. 3: Summary of the cinetic constants, Langmuir, Freundlich, and Temkin parameters for the adsorption of methylene blue in the presence of nanoparticles prepared from laurel leaves.

Kinetic equations	Constants	Dye concentration		Isotherms	Parameters	Temperature (°C)		
Pseudo first order		20 mg/L	50 mg/L			22	40	55
	K1 (min)$^{-1}$	0.04	0.05		q_m(mg.g)$^{-1}$	125	111	111.11
	q(mg.g)$^{-1}$	8.241	21.627	Langmuir	K_L(L. g)$^{-1}$	0.007	0.011	0.016
	R^2	0.819	0.811		R^2	0.96	0.99	0.99
Pseudo-second order	K_2	0.0141	0.004	Parameters thermodynamics	$\Delta H°$ (KJ mol)$^{-1}$		-20.828	
	q	16.66	41.666		$\Delta S°$ (J mol^{-})		-27.544	
	h	3.937	7.692		$\Delta G°$ (KJ mol)$^{-1}$	12.202	11.706	11.293
	R^2	0.996	0.993	Freundlich	K_F(L.g)$^{-1}$	1.196	2.59	5.38
Elovich	o(mg.g^{-1} .min)$^{-1}$	11.326	16.548		n	1.29	1.577	1.968
	β (mg.g^{-1} .min)$^{-1}$	0.328	0.119		R^2	0.94	0.96	0.97
	R^2	0.903	0.881		B	20.64	20.64	20.64
Intra-particle diffusion	K(mg.g^{-1} .min)$^{1/2}$	4.899	1.981	Temkin	A_{tL-gT})$^{-1}$	0.58	0.55	0.498
	R^2	0.854	0.842		R^2	0.95	0.95	0.95

Table III. 4: Summary of the cinetic constants, Langmuir, Freundlich, Temkin, and Dubinin parameters for the adsorption of methylene blue in the presence of nanoparticles prepared from malt fungi.

nanoparticles prepared from malt fungi.

	Cinetic data				Isothermal data			
	Constants	Concentration of dye (mg/L)		Isotherms	Parameters	Temperature (°C)		
Equations cinetics								
First username		25	50			22	40	55
order	K1 (min)$^{-1}$	0.034	0.035	Langmuir	q_L(mg.g)$^{-1}$	125	111.11	111.11
	q_e(mg.g)$^{-1}$	1.63	8.41		K_L (L. g)$^{-1}$	0.0032	0.005	0.007
	R^2	0.89	0.92		R^2	0.87	0.95	0.98
Pseudo-second	K_2	0.066	0.004	Parameters	$\Delta H°$ (KJ mol)$^{-1}$		-19.07	

Order	q_e	5.35	16.39		thermodynamics	AS° (J mol)$^{-1}$		-16.91	
	h	4.69	1.08			AG° (KJ mol)$^{-1}$	14.08	13.78	13.52
R^2		1	0.99			$_{KF}$(L.g)$^{-1}$	2.08	1.021	1.82
Elovich	a	849187	15.42	Freundlich	N	1.16	1.34	1.52
	$в$	3.57	0.36		R^2	0.95	0.95	0.95
	$_{R}2$	0.93	0.89	Temkin	B (j/mol)	19.23	19.65	20.55
Broadcasts	$_{Kid}$	0.57	1.94		A(L.g)$^{-1}$	0.069	0.084	0.098
intra-particulate	(mg.g^1 .min)$^{1/2}$				$_{bt}$ (j/mol)	127.54		
							132.43	132.7
	$_{R}2$	0.61	0.82		$_{R}2$	0.94	0.97	0.97
				Dubinin-Radushkevich	$_{qDR}$ (mg.g)$^{-1}$	38.09	45.56	51.62
					$_{KDR}$ (mol /kJ^{22}) E	5x10^{-5}	4x10^{-5}	3x10^{-5}
					(KJ/mol)	100	111.8	129.1
					R2	0.61	0.73	0.73

The theoretical isotherms of Langmuir, Freundlich, Temkin and Dubinin were used to understand the adsorption mechanism of methylene blue in the presence of prepared copper oxide nanoparticles. The corresponding parameters were determined and summarised in Tables III.2-4. The results obtained show that the Temkin isotherm provides a better fit to the methylene blue adsorption data with high regression coefficients (0.94 $<R^2$) in the case of nanoparticles prepared from pergularia leaves and malt fungi. This implies that the binding energy of the adsorption process decreases linearly with increasing surface coverage [110]. In the case of nanoparticles prepared from bay leaves, the adsorption process follows the Langmuir model. According to Dubinin's equation, the calculated free energy values indicate that the mechanism is chemical. The values of the constant n, determined from the Freundlich equation, are between 1 and 10, which means that the adsorption is favourable.

Negative values of AS° indicate a decrease in disorder at the solution interface. Positive values of AG° suggest the non-spontaneity of the reaction. Negative enthalpy values indicate that the interaction between nanoparticles prepared from pergularia leaves and methylene blue is exothermic. The positive values of AH° in the case of nanoparticles prepared from laurel leaves and malted mushrooms prove an endothermic process. These results are in good agreement with the calculated binding energy values and the adsorbed quantities of dye molecules recorded at different temperatures.

II.5. Catalytic reduction of methylene blue

We are interested here in studying the catalytic power of the degradation of methylene blue in the presence of NaBH4 using nanoparticles prepared from pergularia leaves as an example. The results, presented in Figure III.21, indicate that the copper nanoparticles prepared can be considered as good catalysts for the treatment of methylene blue in an aqueous medium in the presence of $_{NaBH4as}$ a reducing agent. Indeed, the catalytic yield is greater than 99% after only 2 minutes of reaction under the following conditions; pH = 6, NaBH4 = 0.0057 mg, c0 = 10 mg/L, and catalyst dose = 0.005 g (Figure III.21a, c). Also, the solution is completely discoloured after 10 minutes of reaction for a methylene blue concentration equal to 20 mg/L (Figure III.21b,c). The rate constants, determined graphically by plotting Ln Ct/Co against time, are equal to 0.462 and 1.157 min^{-1} , respectively, for methylene blue concentrations of 10 and 20 mg/L.

Table III.5 gives a summary of some methylene blue catalytic reduction results for certain nanomaterials studied in the literature [111-115]. Referring to these results, we note that our prepared catalyst remains excellent and competitive. On the basis of the adsorption capacities recorded and the catalytic yield obtained from the synthesised copper nanoparticles, the results proved the effectiveness of using these copper oxide nanoparticles in both the adsorption and catalytic fields.

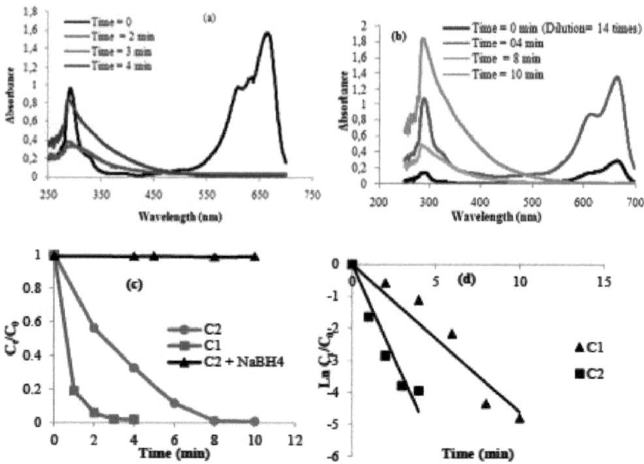

Figure III. 21: (a) Evolution of absorbance as a function of time (C0 = 10 mg/L, pH = 6, T = 22°C), (b) (C0 = 20 mg/L, pH = 6, T = 22°C), (c) variation of Ct/C0 as a function of time, and

(d) variation of Ln Ct/C0 as a function of time.

Table III. 5: Comparison of methylene blue reduction capacities for certain catalysts studied in the literature.

catalysts studied in the literature.

Catalyst	Quantity of catalyst	NaBH4	Methylene blue	Speed constant, k (min^{-1})	Reaction time (min)	Reference
CuO prepared from pergularia leaves	0.005 g	0.0057 mg	10 mg/L	0.462	2	This study
CuO prepared from pergularia leaves	0.005g	0.0057 mg	20 mg/L	1.157	8	This study
Ag/rGO	10 mg	30 mg	50 mL (20 mg/L)	0.155	30	[278]
AgNPs	--	2 m L (0.2 M)	50 mL (10 mg/L)	0.137	14	[279]
Fe3O4@HA@Ag MNCs	1 mg	1 mL (100 mM)	10 mM (100 pL)	0.08	20	[280]
Ag/PSNM-3	2 mg	0.5 mL (60 mM)	2 mL (0.03 mM)	0.134	11	[281]
Ag/GERD	0.25 mg	8 mg	80 mg (0.05 mg/mL)	0.014	---	[282]

Conclusion

To summarise, a ëcological synthesis process was proposedëe to prepare mëtalllcic nanoparticles using lignin extracted from poplar fibres, biological extracts from pergularia tomentosa leaves, oleander leaves and malt fungi. The synthesis of the nanoparticles was confirmed by IR, SEM, XRD and TEM analyses. The maximum adsorption capacities of methylene blue reached 93.2 mg/g, 81.2 mg/g and 64 mg/g, respectively, for nanoparticles prepared from pergularia leaves, bay leaves and malted mushrooms. The adsorption phenomenon was best described by the pseudo-second-order kinetic equation indicating a chemical mechanism. Nanoparticles prepared from pergularia leaves showed excellent methylene blue reduction efficiency in the presence of NaBH4. After a reaction period of just 02 minutes, the coloured solution was completely decoloured (pH = 6, NaBH4 = 0.0057 mg, C0 = 10 mg/L). These interesting results proved that the nanostructures prepared could be put to a wide range of uses.

General conclusion and outlook

The work presented in this accreditation aims to develop the use and development of new biomass-based materials that are abundant, recyclable and less expensive. In parallel with this main objective, particular attention has been paid to local marine and vëgëtal biomass.

Throughout this study, we have focused on the study of adsorption procës, one of the procëdës commonly employed, to ë assess the adsorption ctipticities of raw and also functionalized agricultural and marine wastes vis-a-vis textile dyes.

In the second chapter, we focused on the exploitation and characterisation of new materials derived from shrimp shells, almond fruit waste, plant fibres and date waste. The functionalization of chitosan with amino silica and the other cellulosic materials with either the chitosan polymer or a quaternary copolymer compensated for the low affinity of the cellulosic materials for anionic dyes. All the materials showed very competitive adsorption performances for different classes of dyes: cationic, acid, reactive and direct.

In the third chapter, we proposed new dyestuffs derived from walnut and malt fungi for dyeing textile materials. The results showed the importance of the products studied in terms of chemical composition and dyeing performance. An easy and environmentally friendly route was also developed, for the first time, to provide antimicrobial printed cellulose materials. IR, SEM and XRD spectroscopy results confirmed the success of the method in providing a print. Colour strength depends on the nature of the metal and its concentration. Wash fastness and rub fastness are excellent. The biological activities of fabrics printed with the three types of nanoparticles studied showed their ability to reduce the total viable number of strains of *Staphylococcus aureus*, *Salmonella typhi* and *Candida albicans* in their microbial suspensions. In brief, the design and synthesis thus proposed in this study could create new materials for extending innovative green applications. As this technique can exhibit multiple propriëtës without using auxiliary chemicals, it could be consideredërëe as a dëfi.

In the final chapter, a ëcological svnthesis procës has ël ë proposëe to prepare copper oxide nanoparticles using lignin extracted from poplar fibres and biological extracts from pergularia tomentosa leaves, oleander leaves and malta fungi. The analysis of the nanoparticles was confirmed using different cara^risation techniquesZR, SEM, DRX, ATG/DTG, and TEM. The maximum iidsorlated quant^s of mëthvlëne blue reached 93.2 mg/g, 81.2 mg/g, and 64 mg/g, respectively, for copper oxide nanoparticles prepared from pergularia leaves, bay leaves, and malt mushrooms. The nanoparticles produced from pergularia leaves were able to rapidly reduce mëthvlëne blue in the presence of NaBH4 with yields reaching 100% in some cases. As an example, after a reaction përiode of only 02 minutes, the mëthvlëne blue solution had ël ë complement dëcolorëe (pH = 6, NaBH4 = 0.0057 mg, c_o = 10 mg / L).

In short, we have valorisedë several biomatërials and biomolecules for ëcological dyeing and water treatment applications eontiimiated into textile dyes. It turned out that fauna is rich in value-added materials that can be used for human needs such as medicine, environmental and industrial applications, etc. This ëbeing one of the applications of these biomatërials is successfully completed, the future work of this habilitation remains varied and open-ended and we intend to dëvelop new nanocomposite matërials based on these biomatërials ëtudiës. Our study could also be extended to synthetising metallic nanoparticles with other biological extracts and evaluate other uses including printing, reproduction of new colours, study of biological actives against pathogenic agents, treatment of water contaminated with polyphenols, heavy metals, pesticides, etc.

References

References

1.   Calero, M., I'a'nez-Rodriguez, I., P'erez, A., Martin-Lara, M.A., Bl'azquez, G., 2018. Neural fuzzy modelization of copper removal from water by biosorption in fixed-bed columns using olive stone and pinion shell. Bioresour. Technol. 252, 100-109.

2.   Azari, A., Noorisepehr, M., Dehghanifard, E., Karimyan, K., Hashemi, S.Y., Kalhori, E.M., Norouzi, R., Agarwal, S., Gupta, V.K., 2019. Experimental design, modeling and mechanism of cationic dyes biosorption on to magnetic chitosan-glutaraldehyde composite. Int. J. Biol. Macromol. 131, 633-645.

3.   Qadri, R., Faiq, M.A., 2020. Freshwater pollution: effects on aquatic life and human health. In: Qadri, H., Bhat, R.A., Mehmood, M.A., Dar, G.H. (Eds.), Fresh Water Pollution Dynamics and Remediation. Springer Singapore, Singapore, pp. 15-26.

4.   Yin, K., Wang, Q., Lv, M., Chen, L., 2019. Microorganism remediation strategies towards heavy metals. Chem. Eng. J. 360, 1553-1563.

5.   Hou, T., Du, H., Yang, Z., Tian, Z., Shen, S., Shi, Y., Yang, W., Zhang, L., 2019. Flocculation of different types of combined contaminants of antibiotics and heavy metals by thermo- responsive flocculants with various architectures. Separ. Purif. Technol. 223, 123-132.

6.   Chen, Q., Yao, Y., Li, X., Lu, J., Zhou, J., Huang, Z., 2018. Comparison of heavy metal removals from aqueous solutions by chemical precipitation and characteristics of precipitates. J. Water Process Eng. 26, 289-300.

7.   Feng, Y., Yang, S., Xia, L., Wang, Z., Suo, N., Chen, H., Long, Y., Zhou, B., Yu, Y., 2019. In-situ ion exchange electrocatalysis biological coupling (i-IEEBC) for simultaneously enhanced degradation of organic pollutants and heavy metals in electroplating wastewater. J. Hazard Mater. 364, 562-570.

8.   Du, W.-N., Chen, S.-T., 2018. Photo- and chemocatalytic oxidation of dyes in water. J. Environ. Manag. 206, 507-515

9.   Doggaz, A., Attour, A., Le Page Mostefa, M., Come, K., Tlili, M., Lapicque, F., 2019. Removal of heavy metals by electrocoagulation from hydrogenocarbonate-containing waters: compared cases of divalent iron and zinc cations. J. Water Process Eng. 29, 100796.

10. Nemati, M., Hosseini, S.M., Shabanian, M., 2017. Novel electrodialysis cation exchange membrane prepared by 2-acrylamido-2-methylpropane sulfonic acid; heavy metal ions removal. J. Hazard Mater. 337, 90-104.

11. Hosseini, S.A., Vossoughi, M., Mahmoodi, N.M., Sadrzadeh, M., 2018. Efficient dye removal from aqueous solution by high-performance electrospun nanofibrous membranes through incorporation of SiO2 nanoparticles. J. Clean. Prod. 183, 1197-1206.

12. Kavitha, E., Sowmya, A., Prabhakar, S., Jain, P., Surya, R., Rajesh, M.P., 2019. Removal and recovery of heavy metals through size enhanced ultrafiltration using chitosan derivatives and optimization with response surface modeling. Int. J. Biol. Macromol. 132, 278-288.

13. Qiu, M., He, C., 2019. Efficient removal of heavy metal ions by forward osmosis membrane with a polydopamine modified zeolitic imidazolate framework incorporated selective layer. J. Hazard Mater. 367, 339-347.

14. Wei, Y., Salih, K.A.M., Rabie, K., Elwakeel, K.Z., Zayed, Y.E., Hamza, M.F., Guibal, E., 2021. Development of phosphoryl-functionalized algal-PEI beads for the sorption of Nd(III) and Mo(VI) from aqueous solutions - application for rare earth recovery from acid leachates. Chem. Eng. J. 412, 127399.

15. Napoli, M., Cecchi, S., Grassi, C., Baldi, A., Zanchi, C.A., Orlandini, S., 2019.

Phytoextraction of copper from a contaminated soil using arable and vegetable crops. Chemosphere 219, 122-129.

16. Jacob, J.M., Karthik, C., Saratale, R.G., Kumar, S.S., Prabakar, D., Kadirvelu, K., Pugazhendhi, A., 2018. Biological approaches to tackle heavy metal pollution: a survey of literature. J. Environ. Manag. 217, 56-70.

17. Fernando, I.P.S., Sanjeewa, K.K.A., Kim, S.-Y., Lee, J.-S., Jeon, Y.-J., 2018. Reduction of heavy metal ($Pb2+$) biosorption in zebrafish model using alginic acid purified from Ecklonia cava and two of its synthetic derivatives. Int. J. Biol. Macromol. 106, 330-337.

18. Poletto, M.P., Zattera, A.J., Santana, R.M.C., 2012. Structural differences between wood species: Evidence from chemical composition, FTIR spectroscopy, and thermogravimetric analysis. Journal of Applied Polymer Science 126, 336-343.

19. Popescu, C.M., Singurel, G., Popescu, M.C., Vasile, C., Argyropoulos, D.S., Willfor, S., 2009. Vibrational spectroscopoy and X-ray diffraction methods to establish the differences between hardwood and softwood. Carbohydrate Polymers 77, 851-857.

20. Adel, A.M., Abd El-Wahab, Z.H., Ibrahim. A.A., Al-Shemy, M.T., 2010. Characterization of microcrystalline cellulose prepared from lignocellulosic materials. Part I. Acid catalyzed hydrolysis. Bioresource Technology 101, 4446-4455.

21. Bessadok, A., Marais, S., Roudesli, S., Lixon, C., Mëtayer, M., 2008. Influence of chemical modifications on water-sorption and mechanical properties of Agave fibres. Composite Part A, 39, 29-45.

22. (nilcin, i., 2001. Antioxidant activity of eugenol: A structure and activity relationship study. J. Med. Food 14, 975-985.

23. Mohd, S., Luqman, J.R., Shahid, UI., Mohd, N.B., Mohd, S., Mohd A.K., Faqeer, M., 2016. An eco-friendly dyeing of woolen yarn by Terminalia chebula extract with evaluations of kinetic and adsorption characteristics. Journal of Advanced Research 7, 473-482

24. K. Mukunthan, S. Balaji, Cashew apple juice (Anacardium occidentale L.) speeds up the synthesis of silver nanoparticles, Int. J. Green Nanotechnol. 4 (2012) 71-79.

25. H.A. Abd El-Rahman, A.A. El-Badry, O.M. Mahmoud, F.A.Harraz, The effect of the aqueous extract of Cynomorium coccineum on the epididymal sperm pattern of the rat. Phytother. Res. 13 (1999) 248-250.

26. K. Tahir , S. Nazir, B. Li, A. U. Khan, H.K. Zu, P. Yu, S. Gong, S.U. Khan, A. Ahmad, Nerium oleander leaves extract mediated synthesis of gold nanoparticles and its antioxidant activity.Mat. Lett.156 (2015) 198-201.

27. R. Subbaiya, M. Shiyamala, K. Revathi, R. Pushpalatha, M.M. Selvam, Biological Synthesis of Silver Nanoparticles from Nerium oleander and its Antibacterial and Antioxidant Property, Int. J. Current. Microb. App. Sci. 3 (2014) 83-87.

28. P.R. Prasad, S. Kanchi, E.B. Naidoo, In-vitro evaluation of copper nanoparticles cytotoxicity on prostate cancer cell lines and their antioxidant, sensing and catalytic activity: one-pot green approach, J. Photochem. Photobiolog. B: Biolog. 161 (2016) 375-382.

29. S. Boufi, M. Abid, S. Bouattour, A.M. Ferraria, D.S. Conceigao, L.F. Vieira Ferreira, G. Corbel, P.M. Neto, P.A. Lopes, M. Rei Vilar, A.M. Botelho do Rego, Cotton functionalized with nanostructured TiO_2-Ag-AgBr layer for solar photocatalytic degradation of dyes and toxic organophosphates, Int. J. Biol. Macromol. 1 (2019) 902-910.

30. M. Jabli, E. Gamha, N. Sebeia, M. Hamdaoui, Almond shell waste (Prunus dulcis): Functionalization with [dimethy-diallyl-ammonium-chloride-diallylamin-co-polymer] and chitosan polymer and its investigation in dye adsorption, J. Mol. Liq. 240 (2017) 35-44.

31. N. Sebeia, M. Jabli, A. Ghith, Y. Elghoul, F.M Alminderej, Production of cellulose from Aegagropila Linnaei macro-algae: Chemical modification, characterization and application for the bio-sorption of cationic and anionic dyes from water, Int. J. Biol. Macromol. 135 (2019) 152-162.

32. E. Mouxiou, I. Eleftheriadis, N. Nikolaidis, E. Tsatsaroni, Reactive dyeing of cellulosic fibers: Use of cationic surfactants and their interaction with reactive dyes. J. App. Polym. Sci. 108 (2008) 1209-1215.

33. R. Nebojsa, R. Ivanka, Cationic Modification of Cotton Fabrics and Reactive Dyeing Characteristics. J. Eng. Fib. Fab. 7 (2012) 113-121.

34. J.P. Ruparelia, A.K. Chatterjee, S.P. Duttagupta, S. Mukherji, Strain specificity in antimicrobial activity of silver and copper nanoparticles. Actabiomaterialia, 4 (2008) 707-716.

35. S.M. Dizaj, F. Lotfipour, M. B. Jalali, M.H. Zarrintan, K. Adibkia, Antimicrobial activity of the metals and metal oxide nanoparticles, Materials Science and Engineering: C, 44 (2014) 278-284.

36. S.K. Kailasa, T.J. Park, J.V. Rohit, J.R. Koduru, Antimicrobial activity of silver nanoparticles. In Nanoparticles in Pharmacotherapy, William Andrew Publishing (2019) 461484.

37. Lagergren S: Zur theorie der sogenannten adsorption geloster stoffe. Kungliga Svenska Vetenskapsakademiens 1898; Handlingar 24 1-39

38. Y.S. Ho, G. Mc Kay. Process Biochemistry, 34 (1996) 451.

39. A. Ornek, M. Ozacar, I.A. §engil, Biochemical Engineering Journal, 37 (2007) 192.

40. Weber WJ, Morris JC: Kinetics of adsorption on carbon from solution. Journal of the Sanitary Engineering Division 1963;89:31-60.

41. S.J. Allena, G.Mckay,J.F.Porter, Adsorption isotherm models for basic dye adsorption by peat in single and binary component systems, Journal of Colloid and Interface Science, 280, 2004, 322-333.

42. M.I. Temkin, V. Pyzhev, Kinetic of ammonia synthesis on promoted iron catalyst, Acta Physiochim. 12 (1940) 327-356.

43. M.M. Dubinin, The potential theory of adsorption of gases and vapors for adsorbents with energetically non-uniform surface, Chem. Rev. 60 (1960) 235-266.

44. M. Rajabia, K. Mahanpoora, O. Moradib, Preparation of PMMA/GO and PMMA/GO-Fe3O4 nanocomposites for malachite green dye adsorption: Kinetic and thermodynamic studies, Composites Part B 167 (2019) 544-555.

45. Qu, R., Zhang, Y., Qu W., Sun, C., Chen J., Ping Y., Chen H., & Niu,Y. (2013). Mercury adsorption by sulfur- and amidoxime-containing bifunctional silica gel based hybrid materials, Chemical Engineering Journal, 219, 51-61.

46. Wang, T., Turhan, M., Gunasekaran, S. Selected properties of pH-sensitive, biodegradable chitosan-poly(vinyl alcohol) hydrogel. Polymer International, 53 (2004) 911-918.

47. Chiou, M.S., & Li, H.Y, Adsorption behavior of reactive dye in aqueous solution on chemical cross-linked chitosan beads. Chemosphere 50 (2003)1095-1105.

48. V. Vimonses, S. Lei, B. Jin, C.W.K. Chow, C. Saint, Adsorption of congo red by three Australian kaolins, Appl. Clay Sci. 43 (2009) 465-472.

49. A. Gucek, S. Sener, S. Bilgen, A. Mazmanci, Adsorption and kinetic studies of cationic and anionic dyes on pyrophyllite from aqueous solutions. J. Coll. Interf. Sci. 286 (2005) 53-

60.

50. Y.S. Ho, G. McKay, Pseudo-second order model for sorption processes. Process Biochemistry 34 (1999) 451-465.

51. Muthukumaran, C., Sivakumar, V.M., Thirumarimurugan, M., 2016. Adsorption isotherms and kinetic studies of crystal violet dye removal from aqueous solution using surfactant modified magnetic nanoadsorbent. J. Taiwan. Inst. Chem. Eng. 63, 354-362.

52. Aliabadi, R.S., Mahmoodi, N.O., 2018. Synthesis and characterization of polypyrrole, polyaniline nanoparticles and their nanocomposite for removal of azo dyes; sunset yellow and Congo red. J. Clean. Prod. 179, 235-245.

53. Y.S. Ho, G. McKay, The kinetics of sorption of basic dyes from aqueous solution by sphagnum moss peat, Can. J. Chem. Eng. 76 (4) (1998) 822-827.

54. H. Runping, H. Pan, C. Zhaohui, Z. Zhenhui, T. Mingsheng, Kinetics and isotherms of neutral red adsorption on peanut husk. J. Env. Sci. 20 (2008) 1035-1041.

55. Y. Miyah,A. Lahrichi,M. Idrissi, A. Khalil, F. Zerrouq, Adsorption of methylene blue dye from aqueous solutions onto walnut shells powder: Equilibrium and kinetic studies. Surf. Interf.11 (2018) 74-81.

56. J. Mahjoub, M.H.V. Mohamed, S.R. Mohamed, B. Aghleb, Adsorption of Acid Dyes from Aqueous Solution on a Chitosan-cotton Composite Material Prepared by a New Pad-dry Process, J. Eng. Fib. Fab. 6 (2011) 1-12.

57. A.I. Martin, M. Sanchez-Chaves, F. Arranz, Synthesis, characterization and controlled release behaviour of adducts from chloroacetylated cellulose and -naphthylacetic acid. React. Functional Polym. 39 (1999) 179-187.

58. A.M.A. Nada, Hassan, M. L, Thermal Behavior of Cellulose and Some Cellulose Derivatives, Polym. Degrad. Stab. 67 (2000) 111-115.

59. T.A. Saleh, N.M. Tuzen, A. Sari, Polyethylenimine modified activated carbon as novel magnetic adsorbent for the removal of uranium from aqueous solution, Chem. Eng. Research and Design. 117 (2017) 218 -227.

60. I.M. De Rosa, J.M. Kenny, D. Puglia, C. Santulli, F. Sarasini, Morphological, thermal and mechanical characterization of okra (Abelmoschus esculentus) fibres as potential reinforcement in polymer composites, Comp. Sc. Techn. 70 (2010) 116-122.

61. A. Bessadok, S. Marais, S. Roudesli, C. Lixon, M. Mëtayer, Influence of chemical modifications on water-sorption and mechanical properties of Agave fibres, Composite Part A, 39 (2008) 29-45.

62. A.M. Adel, Z.H. Abd El-Wahab, A.A. Ibrahim. M.T. Al-Shemy, Characterization of microcrystalline cellulose prepared from lignocellulosic materials. Part I. Acid catalyzed hydrolysis, Biores. Technol. 101 (2010) 4446-4455.

63. M.P. Poletto, A.J. Zattera, R.M.C. Santana, Structural differences between wood species: Evidence from chemical composition, FTIR spectroscopy, and thermogravimetric analysis, J. App. Polym. Sci. 126 (2012) 336-343.

64. E. Abraham, B. Deepa, L.A. Pothan, M. Jacob, S. Thomas, U. Cvelbar, R. Anandjiwal, Extraction of nanocellulose fibrils from lignocellulosic fibres: A novel approach. Carbohyd. Polym. 86 (2011) 1468-1475.

65. S. Deshuai, Z. Zhongyi, W. Mengling, W. Yude, American Adsorption of Reactive Dyes on Activated Carbon Developed from Enteromorpha prolifera, J. Anal. Chem. 4 (2013) 17-26.

66. N.B. Douissa, S. Dridi-Dhaouadi S, M.F. Mhenni, Spectrophotometric Investigation of

the Interactions between Cationic (C.I. Basic Blue 9) and Anionic (C.I. Acid Blue 25) Dyes in Adsorption onto Extracted Cellulose from Posidonia oceanic, J. Text. Sci. Eng. 6 (2016) 1-9.

67. J. Acharya, J.N. Sahu, C.R. Mohanty, B.C. Meikap, Removal of Lead(II) from Wastewater by Activated Carbon Developed from Tamarind Wood by Zinc Chloride Activation, Chem. Eng. J. 149 (2009) 249-262.

68. A. Gunay, E. Arslankaya, I Tosun, Lead removal from aqueous solution by natural and pretreated clinoptilolite: Adsorption equilibrium and kinetics. J. Hazard. Mater. 146 (2007) 362-371.

69. Y.S. Ho, G. McKay, The kinetics of sorption of basic dyes from aqueous solution by sphagnum moss peat, Can. J. Chem. Eng. 76 (1998), 822-827.

70. S. Richard, M. Boucher, Y. Lalatonne, S. Mëriaux, L. Motte, Iron oxide nanoparticle surface decorated with cRGD peptides for magnetic resonance imaging of brain tumors, Biochimica Biophysica Acta (BBA) - General Subjects, 1861 (2017) 1515-1520.

71. R. Safar, Z. Doumandji, T. Saidou, L. Ferrari,S. Nahle, B. H.Rihn, O. Joubert, Cytotoxicity and global transcriptional responses induced by zinc oxide nanoparticles NM 110 in PMA- differentiated THP-1 cells, Tox. Let308 (2019) 65-73.

72. H. Wang, H. Rao, M. Luo, X. Xue, Z. Xue, X. Lu, Noble metal nanoparticles growth-based colorimetric strategies: From monocolorimetrictomulticolorimetric sensors, Coordination Chemistry Reviews, 398 (2019) 113003.

73. F. Amourizi, K. Dashtian, M. Ghaedi, Polyvinylalcohol-citrate-stabilized gold nanoparticles supported congo red indicator as an optical sensor for selective colorimetric determination of Cr(III) ion, Polyhed176 (2020) 11478.

74. Z. Issaabadi, M. Nasrollahzadeh, S.M. Sajadi, Green synthesis of the copper nanoparticles supported on bentonite and investigation of its catalytic activity. J. Clean. Prod 142 (2017) 35843591.

75. S.B. Reddy, B. K. Mandal, Facile green synthesis of zinc oxide nanoparticles by Eucalyptus globulus and their photocatalytic and antioxidant activity, Adv. Powd. Techn 28 (2017) 785797.

76. K. Saranyaadevi, V. Subha, R.S.E. Ravindran, S. Renganathan, Synthesis and characterization of copper nanoparticle usingCapparis Zeylanica leaf extract, Int. J. Chem. Tech. Res 6 (2014) 4533-4541.

77. B.H. Patel, M.Z. Channiwala, S.B. Chaudhari, A.A. Mandot, Biosynthesis of copper nanoparticles; its characterization andefficacy against human pathogenic bacterium, J. Env. Chem. Eng4 (2016) 2163-2169.

78. J.K.V.M., Angrasan, R. Subbaiya, Biosynthesis of copper nanoparticles by Vitis vinifera leaf aqueous extract and its antibacterial activity, Int. J. Curr. Microbiol. App. Sci 3 (2014) 768774.

79. N. Nagar, V. Devra, Green synthesis and characterization of copper nanoparticles using Azadirachtaindica leaves, Mat. Chem. Phys 213 (2018) 44-51.

80. N. Sebeia, M. Jabli, A. Ghith, Biological synthesis of copper nanoparticles, using Nerium oleander leaves extract: Characterization and study of their interaction with organic dyes, Inorg. Chem. Comm 105 (2019) 36-46.

81. N. Sebeia, M. Jabli, A. Ghith, T.A Saleh, Eco-friendly synthesis of Cynomorium coccineum extract for controlled production of copper nanoparticles for sorption of methylene blue dye, Arab. J. Chem 13 (2020) 4263-4274.

82. M. Jabli, Y.O. Al-Ghamdi, N. Sebeia, S.G. Almalki, W. Alturaiki, J.M. Khaled, A.S. Mubarak, F.K. Algethami, Green synthesis of colloid metal oxide nanoparticles using Cynomorium coccineum: Application for printing cotton and evaluation of the antimicrobial activities, Materials Chemistry and Physics. 249 (2020) 123171.

83. E. Moroydor Derun, N. Tugrul, F. T. Senberber, A. S. Kipcak, S. Piskin. The Optimization of Copper Sulfate and Tincalconite Molar Ratios on the Hydrothermal Synthesis of Copper Borates. World Academy of Science, Engineering and Technology International Journal of Chemical and Molecular Engineering, 10, 2014, 1152-1156.

84. M. Culebras, M. Pishnamazi, G. M. Walker, M. N. Collins. Facile Tailoring of Structures for Controlled Release of Paracetamol from Sustainable Lignin Derived Platforms. Molecules. 26 (2021) 1593.

85. A. Garcia, A. Toledano, M.A. Andres, J. Labidi, Study of the antioxidant capacity of Miscanthus sinensis lignins, Process Biochem. 45 (2010) 935-940.

86. K.R. Aadil, A. Barapatre, S. Sahu, H. Jha, B.N. Tiwary, Free radical scavenging activity and reducing power of Acacia nilotica wood lignin, Int. J. Biol. Macromol. 67 (2014) 220-227.

87. N. Yang, W.H. Li, Mango peel extract mediated novel route for synthesis of silver nanoparticles and antibacterial application of silver nanoparticles loaded onto non-woven fabrics, Ind. Crop. Prod. 48 (2013) 81-88.

88. T.C. Prarthna, N. Chandrasekaran, M. Raichur, A. Mukherjee, Biomimetic synthesis of silver nanoparticles by Citrus limon (lemon) aqueous extract and theoretical prediction of particle. Colloids Surf. B: Biointerfaces. 82 (2011) 152-159.

89. D. Sasidharan, T.R. Namitha, S.P. Johnson, V. Jose, P. Mathew, Synthesis of silver and copper oxide nanoparticles using Myristica fragrans fruit extract: Antimicrobial and catalytic applications, Sustainable Chemistry and Pharmacy. 16 (2020) 100255.

90. T.B. Vidovix, H. B. Quesada, E. F. D. Januario, R. Bergamasco, A. M. S. Vieira, Green synthesis of copper oxide nanoparticles using Punica granatum leaf extract applied to the removal of methylene blue, Materials Letters 257 (2019) 126685.

91. S. Chaudhary, D. Rohilla, A. Umar, N. Kaur, A. Shanavas, Synthesis and characterizations of luminescent copper oxide nanoparticles: Toxicological profiling and sensing applications, Ceramics International. 45 (2019) 15025-15035.

92. A. Nezamzadeh-Ejhieh, S. Hushmandrad. Solar photodecolorization of methylene blue by CuO/X zeolite as a heterogeneous catalyst Applied Catalysis A: General, 388 (2010) 149-159.

93. M. Ramzan, R.M. Obodo, S.Mukhtar, S.Z. Ilyas, F. Aziz, N. Thovhogi, Green synthesis of copper oxide nanoparticles using Cedrus deodara aqueous extract for antibacterial activity, Materials Today: Proceedings. 36 (2020) 576 - 581.

94. A. Nezamzadeh-Ejhieh, N. Moazzeni, Sunlight photodecolorization of a mixture of Methyl Orange and Bromocresol Green by CuS incorporated in a clinoptilolite zeolite as a heterogeneous catalyst. J. Ind. Eng. Chem. 19 (2013) 1433-1442.

95. J. Singh, V. Kumar, K-H. Kim, M. Rawat, Biogenic synthesis of copper oxide nanoparticles using plant extract and its prodigious potential for photocatalytic degradation of dyes, Environ. Res. 177 (2019) 108569.

96. S. Banerjee, M.G. Dastidar. Use of jute processing wastes for treatment of wastewater contaminated with dye and other organics. Bioresour. Technol. 96 (2005) 1919-1928.

97. G. Annadurai, R.-S. Juang, D.-J. Lee, Use of cellulose-based wastes for adsorption of

dyes from aqueous solutions. J. Hazard. Mater. 92 (2002) 263-274.

98. R. Soury, M. Jabli,S. Latif,K. M. Alenezi,M. El Oudi,F. Abdulaziz,Sa. Teka,H. El Moll,A. Haque. Synthesis and characterization of a new meso-tetrakis (2,4,6-trimethylphenyl) porphyrinto) zinc(II) supported sodium alginate gel beads for improved adsorption of methylene blue dye. International Journal of Biological Macromolecules. 202 (2022) 161-176.

99. Almughamisi, M.S., Khan, Z.A., Alshitari, W., Elwakeel, K.Z., 2020. Recovery of chromium(VI) oxyanions from aqueous solution using Cu(OH)2 and CuO embedded chitosan adsorbents. J. Polym. Environ. 28 (1), 47-60

100.Lei, S., Shi, Y., Qiu, Y., Che, L., Xue, C., 2019. Performance and mechanisms of emerging animal-derived biochars for immobilization of heavy metals. Sci. Total Environ. 646, 12811289.

101.B. Chen, C.W. Hui, G. McKay, Film-pore diffusion modeling and contact time optimization for the adsorption of dyestuffs on pith, Chem. Eng. J. 84 (2) (2001) 77-94.

102.G. Skodras, I. Diamantopoulou, A. Zabaniotou, G. Stavropoulos, G.P. Sakellaropoulos, Enhanced mercury adsorption in activated carbons from biomass materials and waste tires, Fuel Process. Technol. 88 (8) (2007) 749-758.

103.A.R. Hidayu, N. Muda, Preparation and characterization of impregnated activated carbon from palm kernel shell and coconut shell for Co2 capture, Procedia Eng. 148 (2016) 106-113.

104.M. Zubair, M. Daud, G. McKay, F. Shehzad, M.A. Al-Harthi, Recent progress in layered double hydroxides (ldh)-containing hybrids as adsorbents for water remediation, Appl. Clay Sci. 143 (2017) 279-292.

105.R. Yuvakkumar, J. Nathanael, S.I. Honga, Rambutan (Nephelium lappaceum L.) peel extract assisted biomimetic synthesis of nickel oxide nanocrystals, Mat. Let 128 (2014) 170174.

106.S. Heneidak, R.J. Grayer, G.C. Kite, M.S.J. Simmonds, Flavonoid glycosides from Egyptian species of the tribe Asclepiadeae (Apocynaceae, subfamily Asclepiadoideae), Biochem. Systemat. Ecol 34 (2006) 575-584.

107.Siham L, Saida O, Moha T, Nadia S, Hakima A (2014) chemical analysis and antioxidant activity of Nerium oleander leaves. OnLine Journal of Biological Science, 14: 1-7.

108.M. Hao-Cong, W. Shuo, L. Ying, K. Yuan-Yuan, M. Chao-Mei, Chemical constituents and pharmacologic actions ofCynomorium plants, Chinese J.Nat. Med.11(2013) 321-329.

109.S. Yallappa, J. Manjanna, M.A. Sindhe, N.D. Satyanarayan, S.N. Pramod, K. Nagaraja, Microwave assisted rapid synthesis and biological evaluation of stable copper nanoparticles using T. arjuna bark extract, Spectrochimica Acta Part A: Mol. Biomol. Spect 110 (2013) 108115.

110.M. Yousefi, S.M. Arami, H. Takallo, M. Hosseini, M. Radfard, H. Soleimani, A.A. Mohammadi, Modification of pumice with hcl and naoh enhancing its fluoride adsorption capacity: kinetic and isotherm studies, Hum. Ecol. Risk Assess. 25 (6) (2019) 1508-1520.

111.Shirsath SR, Patil AP, Patil R, et al. Removal of brilliant greenfrom wastewater using conventional and ultrasonically prepared poly(acrylic acid) hydrogel loaded with kaolin clay: acomparative study. Ultrason Sonochem 2014; 20(3): 914-923.

112.Makhado E, Pandey S, and Ramontja J. Microwave assistedsynthesis of xanthan gum-cl-poly (acrylic acid) based reducedgraphene oxide hydrogel composite for adsorptionof methylene blue and methyl violet from aqueous solution. Int J Biol Macromol 2018; 119: 255269.

113.Garg S and Garg A. Hydrogel: classification, properties, preparation and technical features. Asian J Biomater Res2016; 2(6): 163-170.

114.Ahmed EM. Hydrogel: preparation, characterization, and applications: a review. J Adv Res 2015; 6(2): 105-121.

115.Liang Y, Zhao X, Ma PX. pH-responsive injectablehydrogels with mucosal adhesiveness based on chitosangrafted-dihydrocaffeic acid and oxidized pullulan for localizeddrug delivery. J Colloid Interf Sci 2019; 536: 224-234.

116.

Printed by Books on Demand GmbH, Norderstedt / Germany